SIMPLIFIED DESIGN
OF BUILDING STRUCTURES

Other titles in the
PARKER–AMBROSE SERIES OF SIMPLIFIED DESIGN GUIDES

Harry Parker and James Ambrose
Simplified Design of Concrete Structures, 6th Edition

Harry Parker, John W. MacGuire and James Ambrose
Simplified Site Engineering, 2nd Edition

James Ambrose
Simplified Design of Building Foundations, 2nd Edition

James Ambrose and Dimitry Vergun
Simplified Building Design for Wind and Earthquake Forces, 3rd Edition

Harry Parker and James Ambrose
Simplified Design of Steel Structures, 6th Edition

James Ambrose
Simplified Design of Masonry Structures

James Ambrose and Peter D. Brandow
Simplified Site Design

Harry Parker and James Ambrose
Simplified Mechanics and Strength of Materials, 5th Edition

Marc Schiler
Simplified Design of Building Lighting

James Patterson
Simplified Design for Building Fire Safety

James Ambrose
Simplified Engineering for Architects and Builders, 8th Edition

William Bobenhausen
Simplified Design of HVAC Systems

James Ambrose
Simplified Design of Wood Structures, 5th Edition

James Ambrose and Jeffrey E. Ollswang
Simplified Design for Building Sound Control

SIMPLIFIED DESIGN
OF BUILDING
STRUCTURES

Third Edition

JAMES AMBROSE

Professor of Architecture
University of Southern California
Los Angeles, California

JOHN WILEY & SONS, INC.

New York • Chichester • Brisbane • Toronto • Singapore

Library of Congress Cataloging-in-Publication Data:

Ambrose, James E.
    Simplified design of building structures / James Ambrose—3rd
ed.
        p.   cm.—(Parker Ambrose series of simplified design guides)
    "A Wiley-Interscience publication."
    Includes bibliographical references and index.
    ISBN 0-471-03744-3 (cloth : alk. paper)
    1. Structural design.   2. Architectural design.   3. Buildings.
I. Title. II. Series.
TA658.A5   1995
690—dc20   95-16411

CONTENTS

PREFACE

This new edition of my 1979 book supplements the other volumes in the *Simplified* series on design of building structures. Because of their intentionally compact nature, those books offer more focused views of particular topics, such as applied mechanics, wood structures, steel structures, and so on. This book, meanwhile, looks at structural design from the problem of developing a whole structural system for a defined architectural situation. That is, I describe how to design a system for a proposed building plan, when the building's general form and usage is predetermined.

Professional designers (that is, architects and/or structural engineers) break down structural design into three tasks: consider alternative structural systems and materials, develop the general form and basic elements of a specific system, and, ultimately, define the structure through construction plans and details. Designers thus start with the whole problem and proceed to the complete design in all its necessary parts—all the while keeping in view the concept of the whole building design. Individual design tasks may be few or numerous, simple or complex, highly familiar or totally new. The purpose of the work, however, always is to produce a buildable solution, communicated to the builders in construction documents.

The presentations in this book try to simulate the working design context. In keeping with the overall *simplified* mode, the building problems are relatively simple and of common form, and the structural solutions use highly recognizable materials and elements. The documentation of the design work mostly takes the form of plans and details similar to actual construction documents in current practice in the United States. This book is not intended as a drafting reference, but I have tried to use graphic elements that resemble the form used for construction work at this time.

Some structural computations for the design work are presented here, generally in relatively compact and abbreviated form. I assume that you either have some experience in general structural design or, at the least, have access to other references for fuller explanations of the computational work. Specifically, I assume that you have a general background of involvement represented by the work presented in *Simplified Engineering for Architects and Builders*, 8th edition (Ref. 1). Computational work here that can use that book as a direct reference is presented with little explanation. Work here that goes beyond that book is developed further; I may, in fact, simply refer to some other volume in the Parker/Ambrose series of simplified books (see listing in the front of this book).

Structural design done with imagination and innovation by skilled designers produces inspiring results. But the professional designer is often challenged to visualize the simplest, most direct, most practical solution to a given problem. The ability to produce such solutions is a highly marketable skill. And practical solutions are often a launching point for creative explorations.

This book is intended both for independent study by less experienced structural designers and for classroom use in courses on design of building structures. For the independent reader, I provide references to sources that are commonly available. For the teachers and students, I offer some suggestions for study aids at the back of the book.

Most of all, this book is intended for inexperienced readers. If you later become a professional designer, this book will then seem as trivial as training wheels on a bicycle. It is a starting point, but nowhere near the finish line.

I am grateful to many people for their contributions to my own learning process as a structural designer, a teacher, and a writer of technical books. I am also grateful to the many people at John Wiley and Sons who have helped to make this book possible and shape it into a final, presentable form. I am particularly indebted to my publisher, Peggy Burns; my editors, Everett Smethurst and Amanda Miller; and Robert J. Fletcher IV and Milagros Torres of the Wiley Production division, who helped turn my rough work into a real book.

Finally, once more, I must acknowledge the considerable contributions made by my wife, Peggy, who—aside from her roles as wife and mother—serves many tasks as my professional assistant.

JAMES AMBROSE

Westlake Village, California
1995

SIMPLIFIED DESIGN
OF BUILDING STRUCTURES

INTRODUCTION

Designing building structures involves the consideration of a wide range of factors. Building structural designers must not only understand structural behavior and how to provide for it adequately, but also be knowledgeable about building construction materials and processes, building codes and standards, and building economics. Making sure that the structure is safe is a major professional responsibility, but making sure that the structure gets built is a pressing practical matter.

Building structural designers also must consider the relationship of the building structure to the building as a whole. The structure is one of many subsystems in the building and must coexist with the subsystems for electrical power and lighting, plumbing, heating, vertical transportation, and so on. Designers must ensure that the basic needs of all the building subsystems are met; they must take particular care with those subsystems that require significant space, involve penetrations of the building construction, or relate strongly to the building form. Any building structure is usually compromised to some degree by these concerns.

Structural designers usually come from either an architectural or civil engineering background. Unfortunately, neither of these formal education programs provides a thorough preparation for design work, especially given the broad scope of concerns required for the design of building structures. Architects typically receive too little preparation in engineering investigation and mathematical analysis, while engineers typically get little experience in design work of a broad-scope form and little exposure to general planning or construction of buildings.

1

Aspiring designers must make up for their lack of educational background through independent study or office work experience. Unfortunately, independent study requires learning resources, and few books treat broad-scope design for the inexperienced, engineering analysis for the under prepared, or building planning and design for the uninformed. Moreover, work experience often is narrowly focused on specific project needs and does not develop a general understanding of fundamentals and processes.

This book seeks to bridge some gaps in the backgrounds of potential structural designers. It cannot be a magic cure-all, but it can be useful to the partially prepared reader.

The examples in this book illustrate the problems and processes of designing whole structural systems for buildings. In the examples I first present a general building form, size, and usage as given conditions. (I have selected a range of building uses, sizes, and shapes to give as wide a set of situations as possible for the development of structural solutions.) I then discuss the alternatives for achieving the building construction in general and the structure in particular. This discussion is followed by an illustration of the design of one or more of the viable structural systems. In addition, simple computations are shown for some typical elements of the structural systems. I conclude the example with a presentation of typical construction plans and details for the systems as designed.

I assume that the reader has a minimum background in fundamental mechanics, structural behavior of materials, and elementary investigation of behavior of simple structures. I also assume a reasonable familiarity with the existing practices in building construction with the common structural materials: wood, steel, concrete, and masonry. Readers who are not fully prepared in any of these topics may want to refer to any of several texts in simplified engineering and building construction. This book's bibliography lists several such publications, although others exist.

UNITS OF MEASUREMENT

The work in this book is presented primarily in English units (feet, pounds, and so on). The construction industry in the United States still uses these units, which are now more appropriately called *U.S. units* since England no longer uses them officially. Although the U.S. is gradually switching to SI units, it is hard to estimate how long Americans will hold on to the wood 2 × 4, the 16 in. concrete block, and the 4 ft × 8 ft plywood panel. For some of the computational work here, I give equivalent metric-based data in brackets following the U.S. unit data; I hope these hints help those readers still struggling with the dual-unit situation.

Table 1 lists the standard units of measurement in the U.S. system with the abbreviations used in this book and a description of the units' use in structural work. In similar form, Table 2 gives the corresponding units in the SI system. Table 3 lists the conversion factors used in shifting from one system to the other.

For some of the work in this book, the units of measurement are not significant. What is required in such cases is simply a numerical answer. The visualization of the problem, the manipulation of the mathematical processes for the solution, and the quantification of the answer are not related to the specific units—only to their relative values. In such situations I occasionally do not present the work in dual units, choosing to provide a less confusing illustration.

Note: The structural designer is generally advised to always indicate the units for any numerical answers in structural computations.

COMPUTATIONS

In professional design firms, structural computations are most commonly done with computers, particularly when the work is complex or repetitive. Anyone aspiring to professional design work is advised to acquire the necessary computer skills and experience.

The computational work in this book is simple, however, and can be performed easily with a pocket calculator that has an eight-digit capacity.

For the most part, structural computations can be rounded off; accuracy beyond the third place is seldom significant. In some examples more accuracy is carried in early stages of the computation to ensure the desired degree in the final answer. All the work in this book was performed on an eight-digit pocket calculator.

SYMBOLS

The following "shorthand" symbols are frequently used:

Symbol	Reading
$>$	is greater than
$<$	is less than
$\geqslant$	equal to or greater than
$\leqslant$	equal to or less than
$6'$	six feet
$6''$	six inches
Σ	the sum of
ΔL	change in L

TABLE 1 Units of Measurement: U.S. System

Name of Unit	Abbreviation	Use
Length		
Foot	ft	Large dimensions, building plans, beam spans
Inch	in.	Small dimensions, size of member cross sections
Area		
Square feet	ft^2	Large areas
Square inches	$in.^2$	Small areas, properties of cross sections
Volume		
Cubic feet	ft^3	Large volumes, quantities of materials
Cubic inches	$in.^3$	Small volumes
Force, Mass		
Pound	lb	Specific weight, force, load
Kip	k	1000 pounds
Pounds per foot	lb/ft	Linear load (as on a beam)
Kips per foot	k/ft	Linear load (as on a beam)
Pounds per square foot	lb/ft^2, psf	Distributed load on a surface
Kips per square foot	k/ft^2, ksf	Distributed load on a surface
Pounds per cubic foot	lb/ft^3, pcf	Relative density, weight
Moment		
Foot-pounds	ft-lb	Rotational or bending moment
Inch-pounds	in.-lb	Rotational or bending moment
Kip-feet	k-ft	Rotational or bending moment
Kip-inches	k-in.	Rotational or bending moment
Stress		
Pounds per square foot	lb/ft^2, psf	Soil pressure
Pounds per square inch	lb/in^2, psi	Stresses in structures
Kips per square foot	k/ft^2, ksf	Soil pressure
Kips per square inch	k/in^2, ksi	Stresses in structures
Temperature		
Degree Fahrenheit	°F	Temperature

TABLE 2 Units of Measurement: SI System

Name of Unit	Abbreviation	Use
Length		
Meter	m	Large dimensions, building plans, beam spans
Millimeter	mm	Small dimensions, size of member cross sections
Area		
Square meters	m^2	Large areas
Square millimeters	mm^2	Small areas, properties of cross sections
Volume		
Cubic meters	m^3	Large volumes
Cubic millimeters	mm^3	Small volumes
Mass		
Kilogram	kg	Mass of materials (equivalent to weight in U.S. system)
Kilograms per cubic meter	kg/m^3	Density
Force (Load on Structures)		
Newton	N	Force or load
Kilonewton	kN	1000 newtons
Stress		
Pascal	Pa	Stress or pressure (1 pascal = $1 N/m^2$)
Kilopascal	kPa	1000 pascals
Megapascal	MPa	1,000,000 pascals
Gigapascal	GPa	1,000,000,000 pascals
Temperature		
Degree Celsius	°C	Temperature

TABLE 3 Factors for Conversion of Units

To Convert from U.S. Units to SI Units Multiply by	U.S. Unit	SI Unit	To Convert from SI Units to U.S. Units Multiply by
25.4	in.	mm	0.03937
0.3048	ft	m	3.281
645.2	in.^{2}	mm^2	1.550×10^{-3}
16.39×10^3	in.^{3}	mm^3	61.02×10^{-6}
416.2×10^3	in.^{4}	mm^4	2.403×10^{-6}
0.09290	ft^2	m^2	10.76
0.02832	ft^3	m^3	35.31
0.4536	lb (mass)	kg	2.205
4.448	lb (force)	N	0.2248
4.448	kip (force)	kN	0.2248
1.356	ft-lb (moment)	N-m	0.7376
1.356	kip-ft (moment)	kN-m	0.7376
1.488	lb/ft (mass)	kg/m	0.6720
14.59	lb/ft (load)	N/m	0.06853
14.59	kips/ft (load)	kN/m	0.06853
6.895	psi (stress)	kPa	0.1450
6.895	ksi (stress)	MPa	0.1450
0.04788	psf (load or pressure)	kPa	20.95
47.88	ksf (load or pressure	kPa	0.02093
$0.566 \times (°F - 32)$	°F	°C	$(1.8 \times °C) \times 32$

NOTATION

The use of notation is complicated by the fact that there is a lack of consistency in the notation currently used by structural designers. Some of the standards used in the field are developed by individual groups (notably those relating to a single basic material, such as wood, steel, concrete, masonry, and so on); each group has its own particular notation. Thus the same type of stress (for example, shear stress in a beam) or the same symbol (f_c) may have different representations in structural computations. To keep some form of consistency in this book, I use the following notation:

a	(1) Moment arm; (2) increment of an area
A	Gross (total) area of a surface or cross section
b	Width of a beam cross section
B	Bending coefficient
c	Distance from neutral axis to edge of a beam cross section
d	Depth of a beam cross section or overall depth (height) of a truss
D	(1) Diameter; (2) deflection

e	(1) Eccentricity (dimension of the mislocation of a load resultant from the neutral axis, centroid, or simple center of the loaded object); (2) elongation
E	Modulus of elasticity (ratio of unit stress to the accompanying unit strain)
f	Computed unit stress
F	(1) Force; (2) allowable unit stress
g	Acceleration due to gravity
G	Shear modulus of elasticity
h	Height
H	Horizontal component of a force
I	Moment of inertia (second moment of an area about an axis in the plane of the area)
J	Torsional (polar) moment of inertia
K	Effective length factor for slenderness (of a column: KL/r)
M	Moment
n	Modular ratio (of the moduli of elasticity of two different materials)
N	Number of
p	(1) Percent; (2) unit pressure
P	Concentrated load (force at a point)
r	Radius of gyration of a cross section
R	Radius (of a circle, for example)
s	(1) Center-to-center spacing of a set of objects; (2) distance of travel (displacement) of a moving object; (3) strain or unit deformation
t	(1) Thickness; (2) time
T	(1) Temperature; (2) torsional moment
V	(1) Gross (total) shear force; (2) vertical component of a force
w	(1) Width; (2) unit of a uniformly distributed load on a beam
W	(1) Gross (total) value of a uniformly distributed load on a beam; (2) gross (total) weight of an object
Δ (delta)	Change of
Σ (sigma)	Sum of
θ (theta)	Angle
μ (mu)	Coefficient of friction
ϕ (phi)	Angle

PART I

BUILDING ONE

1

GENERAL CONSIDERATIONS
FOR BUILDING ONE

1.1 THE BUILDING

Building One is a split-level, single-family residence; it's an all-American suburban house (see Figure 1.1).

The light wood frame is predominant for this type of building in all parts of the United States. It remains popular, thanks to the resilience of the timber industry, the continued availability of timber stands in North America, and the highly developed means for timber distribution. Recent surges in cost of lumber and plywood notwithstanding, the basic form of the light wood frame endures—even when executed with wood fiber or metal parts.

Exterior and interior cosmetics vary according to popular styles, while roofing and exterior finishes, window construction, insulation, and weather tightness of enclosure systems depend on local and regional concerns. What does not vary much over time or region is the basic structure that uses studs, joists, rafters, sheathing, and decking.

The construction shown here is mostly typical, favoring a mild climate in some ways. With minor modifications, however, the building shown here could be built just about anywhere in the United States. Notes on Figures 1.2 through 1.5 identify basic elements of the construction and, in some cases, also mention basic issues and possible variations.

BUILDING 1
1 – Living Room
2 – Dining Room
3 – Kitchen
4 – Bedroom
5 – Family Room
6 – Garage

South Elevation

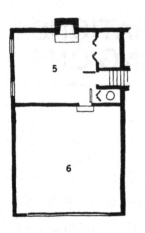

Plan – Lower Level

North

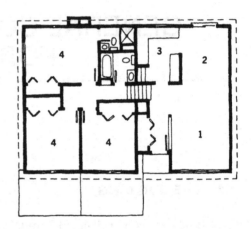

Plan – Upper Level

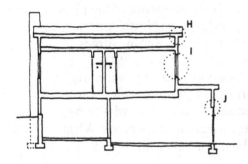

North – South Section

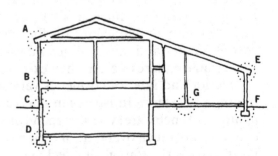

East – West Section

FIGURE 1.1 Building One: general form.

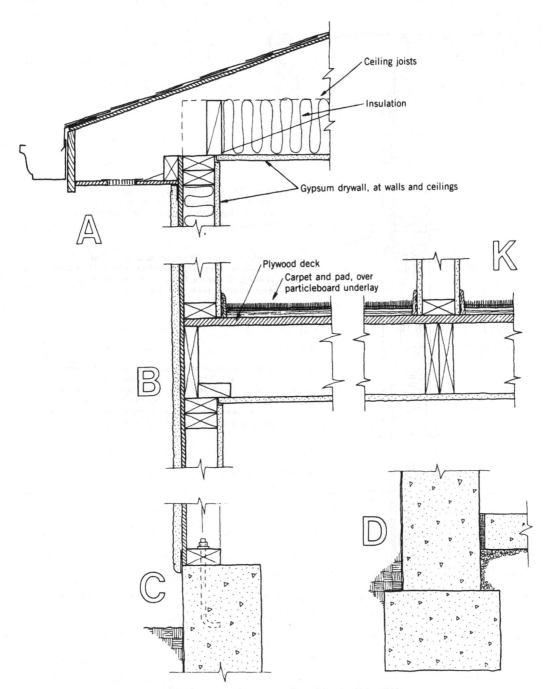

Ceiling joists

Insulation

Gypsum drywall, at walls and ceilings

Plywood deck

Carpet and pad, over particleboard underlay

A

B

K

C

D

FIGURE 1.2 Building One: construction details.

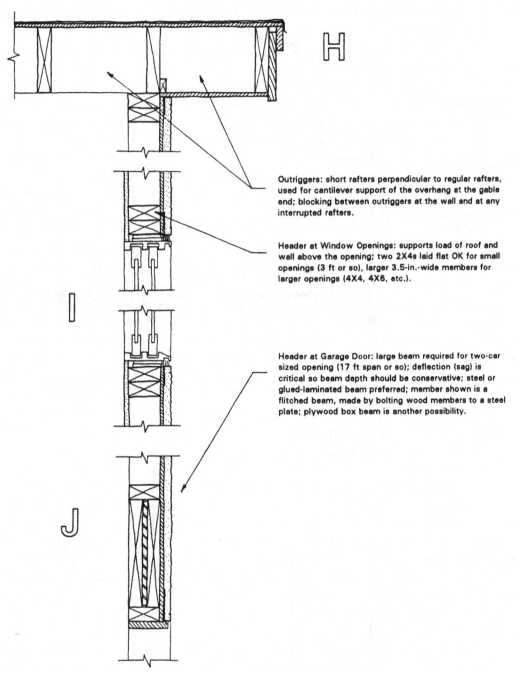

H

I

J

Outriggers: short rafters perpendicular to regular rafters, used for cantilever support of the overhang at the gable end; blocking between outriggers at the wall and at any interrupted rafters.

Header at Window Openings: supports load of roof and wall above the opening; two 2X4s laid flat OK for small openings (3 ft or so), larger 3.5-in.-wide members for larger openings (4X4, 4X6, etc.).

Header at Garage Door: large beam required for two-car sized opening (17 ft span or so); deflection (sag) is critical so beam depth should be conservative; steel or glued-laminated beam preferred; member shown is a flitched beam, made by bolting wood members to a steel plate; plywood box beam is another possibility.

FIGURE 1.3 Building One: construction details.

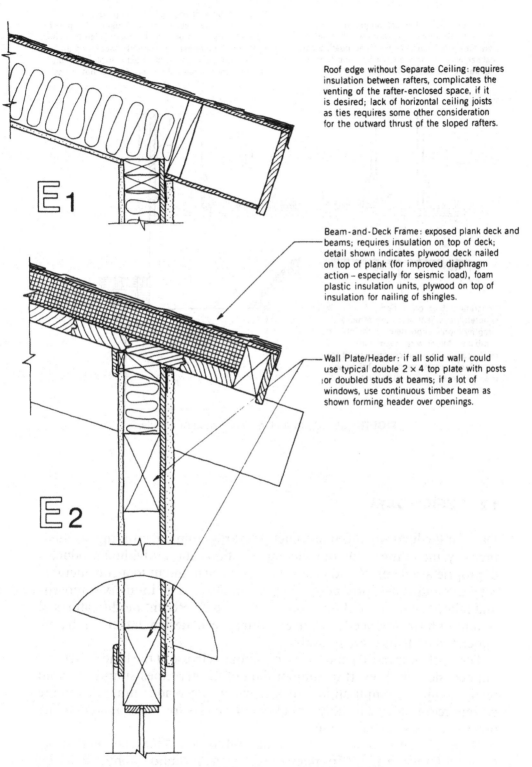

Roof edge without Separate Ceiling: requires insulation between rafters, complicates the venting of the rafter-enclosed space, if it is desired; lack of horizontal ceiling joists as ties requires some other consideration for the outward thrust of the sloped rafters.

Beam-and-Deck Frame: exposed plank deck and beams; requires insulation on top of deck; detail shown indicates plywood deck nailed on top of plank (for improved diaphragm action – especially for seismic load), foam plastic insulation units, plywood on top of insulation for nailing of shingles.

Wall Plate/Header: if all solid wall, could use typical double 2 × 4 top plate with posts or doubled studs at beams; if a lot of windows, use continuous timber beam as shown forming header over openings.

FIGURE 1.4 Building One: construction details.

Stud wall base within slab on grade. Detail shown (with drilled-in anchor for sill) is OK only for a wall that is neither load bearing nor a shear wall. Cast in anchor bolt and footing required for structural wall.

Stud wall base at edge of slab on grade. Wall anchorage essentially same as Detail C. Support for slab edge depends on soil conditions, slab functions, and code requirements. Edge insulation and thermal break required in colder climates. Shallow footing shown here is used only where frost is not critical.

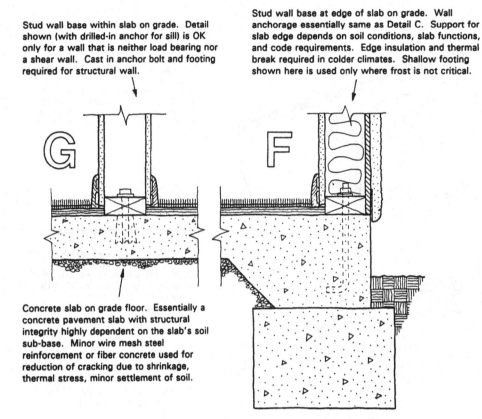

Concrete slab on grade floor. Essentially a concrete pavement slab with structural integrity highly dependent on the slab's soil sub-base. Minor wire mesh steel reinforcement or fiber concrete used for reduction of cracking due to shrinkage, thermal stress, minor settlement of soil.

FIGURE 1.5 Building One: construction details.

1.2 DESIGN DATA

Data for the light wood frame comes primarily from the building products industry, including industry-wide organizations and individual producers of proprietary items. Wood products may be custom-cut to a considerable extent, but the basic products are tightly controlled by industry standards and relate to current building code requirements. Recent modifications of standards have produced a relatively clumsy set of design data that is barely understandable by mere mortals.

The work here mostly conforms to standards in use at the time of writing, but occasionally I revert to simpler data of former times simply to avoid really complex computations. In any event, the reader must determine current requirements for any actual design work—for wood as well as any other materials or design issues.

Design data used here is based on the 1994 edition of the *Uniform Building Code* (UBC), which is the first reference listed in the bibliography. Properties for products are taken from the UBC or from manufacturers' and suppliers' data sources, as noted in the computations.

Regional usage of wood relates to the local availability of trees and lumber industry sources. The products used in the examples in this book are in general use in the western United States. General construction practices and specific product availability will vary for other regions.

1.3 STRUCTURAL ALTERNATIVES

The light wood frame (Figure 1.6) is still predominant. The 2 × 4 stud; the 2 × 6, 8, 10, or 12 joists and rafters; and the 4 ft by 8 ft panel size for sheathing and decking establish a modular system of dimensions. Frame spacings of 12, 16, 24, 32, 48, and 96 in. are thus obvious, and the 3.5-in.-wide void space in walls is typical. You can expect some dimensions to differ from the basic module, but the pressure to reduce waste makes using the readily available units of materials popular. Besides, a major way to cut costs is to reduce the need for on-site labor, particularly of the skilled crafts.

Despite the strong control of the modular dimensions of standard products, this structural system has immense potential for accommodating variations in building form, size, and detail. The materials are relatively easy to work with on- or off-site, and custom forming can be done virtually without limit. This flexibility adds considerably to the enduring popularity of this building type with designers and builders.

Of course, other possible structures work for this type of building. For example, the light wood frame can be emulated in steel, using light-gage elements. Using steel offers the principal advantage of reducing the mass of combustible material in the building. This may be a desirable feature in some situations or an actual building or zoning code requirement in others. Formerly, steel was rarely economically beneficial. With the present cost of lumber, however, the use of light-gage steel frames has increased, although basic system forms generally emulate the good old light wood frame (2 × 4 studs, and so on).

In times past, the wood structure was frequently developed as a heavy timber frame, often with infill utilizing elements of light wood framing. Although the elegantly crafted, exposed wood frame can be very attractive—indeed, dramatic—it is not usually economically feasible for most ordinary buildings. In addition, the lack of available craft workers and the difficulty in obtaining good timber causes problems even when cost is not a concern. Nonetheless, this system currently enjoys a resurgence of popularity.

While the 2 × 4 is the workhorse of the light wood framing system, and the 3.5-in.-wide stud space is most common, wall thickness can vary. For tight planning situations, nonstructural walls of thinner dimension can be produced by using 2 × 3 studs or even 2 × 2 studs. More often, however, there is a compelling need for thicker walls—or more to the point, for a wider void space. The two most common reasons are the need for more insulation in exterior walls and the need to accommodate large items, such as ducts or extensive plumbing.

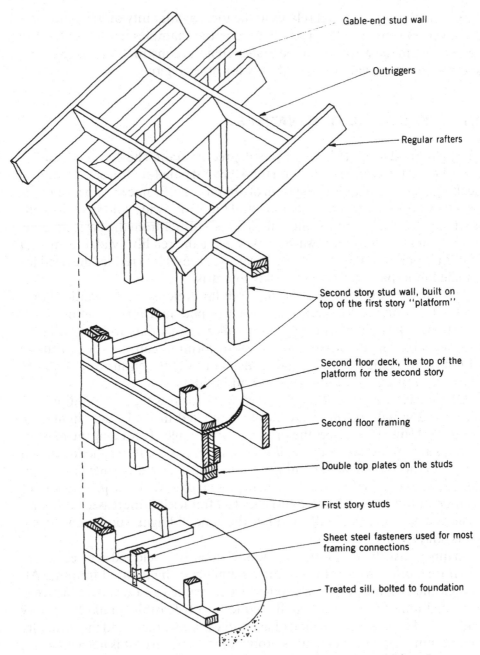

Gable-end stud wall

Outriggers

Regular rafters

Second story stud wall, built on top of the first story "platform"

Second floor deck, the top of the platform for the second story

Second floor framing

Double top plates on the studs

First story studs

Sheet steel fasteners used for most framing connections

Treated sill, bolted to foundation

FIGURE 1.6 Building One: structural drawing of the light wood frame.

The void spaces in walls or between joists and rafters often contain wiring, piping, and ducts, and various other items for the building's subsystems, for electrical power, lighting, water and waste, doorbell, and security alarm systems. The hollow wall space can also contain columns for a heavier frame so that the frame does not intrude in the building occupied space.

Using solid sawn wood pieces for the light wood frame requires a careful control of the quality of the wood used; you must not only be economical, but also minimize the dimensional and shape change due to shrinkage effects in the wood. There has been a general, steady reduction in the quality of lumber used for the light wood frame, mostly for economic reasons. Use of fabricated elements, such as laminated members, wood + plywood I or box sections, and light prefabricated trusses, is becoming more prevalent for floor and roof structures—especially for spans over 15 ft or so.

Wall sheathing and floor and roof decking comprised solid sawn wood boards in the past. Early in this century, though, these boards were gradually replaced, and decking was mostly done with plywood and wall sheathing with plywood and various wood fiber products in panel form. Now compressed-fiber products (particleboard, flakeboard, hardboard, and so on) are steadily replacing plywood for decking, conserving the increasingly scarce high-quality lumber for when it is really needed.

It is possible, of course, to use any form of structure for a house. Steel frames, masonry, reinforced concrete, aluminum skins, and fabric—as well as ice, mud, twigs, and animal skins—have been used. The homemade or high-crafted, custom-designed house offers endless possibilities and some notable, outstanding examples. However, the light wood frame still stands as the most widely used system.

2

THE WOOD STRUCTURE

The building's general form and construction details are shown in Figures 1.2 through 1.6. The construction materials and details for such a building vary considerably because of climate, local building code requirements, and local building practices. There is often little if any actual structural design work done for such a building because of the common, repetitive nature of the construction. The explanations that follow illustrate the nature of the structure, although most of the elements of the system may be obtained from various tabulations in codes or handbooks.

I have for the most part used the *Uniform Building Code* (UBC) as a reference for design criteria, although this code is not really intended to cover single- or two-family housing. As in all cases, consult the local codes before beginning any actual design work.

The materials used for the construction are

Joists, rafters, and studs: No. 2 Douglas Fir–Larch.

Beams and posts: No. 1 Douglas Fir–Larch (4× and wider).

Roof deck, floor deck, and exterior wall sheathing: Douglas Fir plywood, structural grade.

Structural steel: A36, F_y = 36 ksi [248 MPa].

Concrete: stone aggregate, f_c' = 3 ksi [20.7 MPa].

Some of the criteria used for the structural design are

Floor live load: 40 psf [1.92 kPa].

Roof live load: 20 psf (snow) [1.44 kPa].

Wind: 20 psf on vertical surfaces [0.96 kPa].

Soil: 2 ksf maximum [96 kPa].

2.1 THE ROOF STRUCTURE

The roof structure consists of a plywood deck nailed to supporting rafters, roof beams, and tops of the stud bearing walls. A general plan showing the layout of this structure is in Figure 2.1. The 4 in 12, or 1 in 3, roof slope provides slightly longer spans than the horizontal plan dimensions indicate, but the usual practice is to design for the horizontal projection of the inclined rafters; since tables for rafters are typically based on gravity loading only, this practice is generally acceptable.

Choice for the plywood deck relates to four major concerns:

1. *Supports for the deck (in this case, the parallel rafters).* This determines the deck span and some aspects of fitting the deck panels to the support system. Using common 4 ft by 8 ft panels, logical rafter spacings are 12, 16, 24, 32, or 48 in.
2. *Attachment of the roofing.* High-slope roofs usually have nailed shingles or tiles. A minimum deck thickness is required for holding nails, unless nailing is otherwise developed.
3. *Required diaphragm functions of the roof deck for lateral loads of wind or earthquakes.*
4. *Available sizes (thicknesses) of structural plywood panels.* These sizes are now in metric units but still approximate old thicknesses of $\frac{3}{8}$, $\frac{1}{2}$, $\frac{5}{8}$, and $\frac{3}{4}$ in.

The minimum high-slope deck is typically a $\frac{3}{8}$ in. plywood, which is usually adequate for composition or wood shingles. For flat roofs, a $\frac{1}{2}$ in. deck is usually required.

Deck load/span capacities may be obtained from Tables B.5 and B.6. The choice of table depends on the span direction related to the plywood face grain direction.

Building structural designers also must consider the plywood panel edges perpendicular to the supporting rafters, which normally are not backed up with wood framing; as a result, panels can move independently under load, possibly causing fracture of roofing materials. Remedies include using tongue and groove joints (only feasible with panels of $\frac{3}{4}$ in. thickness or greater), providing wood blocking between the rafters for nailing of the otherwise unsupported edges, or holding the two edges together with a metal H-shaped clip.

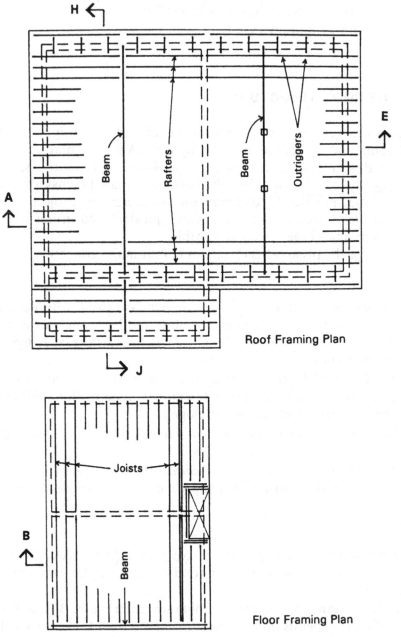

FIGURE 2.1 Framing plans for the roof and upper-level floor.

Rafters

A complete rafter design includes considerations for bending stress and deflection due to the various combinations of dead, live, and wind loads. The wind load is assumed to act normal to the roof surface, but the gravity loads produce bending only as vectors in the direction perpendicular to the rafter span (see Figure 2.2).

For the relatively low-slope rafter and a low wind force, it is common to ignore the wind and to design the rafter on the basis of the gravity loads on the horizontal span. You can find tables of allowable load/span conditions for ordinary rafter sizes of structural lumber in various references. Inspection of one such table, Table B.12, produces the following alternatives (all Douglas Fir–Larch, No. 2 grade):

Span = 12 ft, E = 1,600,000 psi, F_b = 1006 psi
Table choices: 2 × 8s at 16 in. on center, or 2 × 10s at 24 in. on center

(Allowable stress and modulus of elasticity values are from Table B.4.)

Solid-sawn lumber is most practical for this short span. However, for longer spans—those that exceed the capability of 2 × 12 lumber—various other products, including laminated and built-up elements and light wood and steel combination trusses, are available and more practical.

Ceiling Joists

The underside of a high-slope roof is also sloped; it is easy to develop a corresponding sloped ceiling surface, as in the Building One living room and

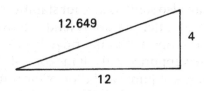

The Rafter Geometry

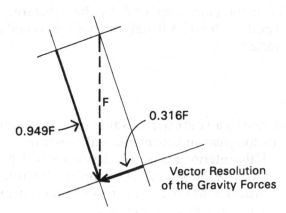

Vector Resolution
of the Gravity Forces

FIGURE 2.2   Resolution of the roof slope.

dining room areas. In this case the ceiling is probably not a separate structure, unless the space between rafters is too great for the panel materials used to develop the ceiling surface. There may, in fact, be no ceiling—that is, no surfacing materials are applied to the underside of the roof structure. (In truth, there is always a ceiling; it is just not always separately developed.) This exposed structure, shown in Figure 1.4, is an option for the roof at the living room.

The drawings indicate the use of a flat ceiling surface in the bedrooms at the level of the roof's low point, thus creating an attic space. To create such a flat ceiling, you must use a separate framing system, to which you can affix the ceiling surface materials. For this example the most direct solution is to add a set of joists, spanning between the outside walls and available interior walls.

Ceiling joists can be quite minimal if it is assumed that the space above them is not occupied. If so, the joists must support only the dead weight of the ceiling; they should have a sufficient stiffness so as to avoid excessive sag. If the attic is occupied, however, then the joists are really floor joists. If the attic is not supposed to be occupied but is large enough to crawl into, it should be assumed that people will succumb to the temptation to use it for storage. Therefore, the ceiling joists should be able to hold the weight of an average person without serious structural challenge.

Consult a table like Table B.10 to select appropriate ceiling joists where loads from above are considered minimal. Table B.10 uses an arbitrary live load of 10 psf for a minimal deflection consideration.

The ceiling joist can also serve as a tie for stability of the rafters. As shown in Figure 2.3, the sloped rafters naturally tend to move outward at their supports unless they are supported at the ridge by a wall or a separate ridge beam. When they are supported only at the tops of the outside walls, some form of lateral bracing is required at the top of the wall. Flying buttresses are a dramatic solution; the most practical remedy is a horizontal tie across the building, making the two rafters work against each other to produce an equilibrium of horizontal forces internally to the structure. The horizontal ceiling joists can perform this task if they are continuous and are adequately attached to the rafters.

Roof Supports

The roof is supported by all the exterior walls and some interior walls. If the spans of rafters are too great and no interior wall exists, the roof is supported by special beams. If the interior is generally developed with partitioning, you probably won't need special beams. If an open planning design is developed, the roof structure may become a major beam system. Designs for a few special beams are presented in Section 2.4.

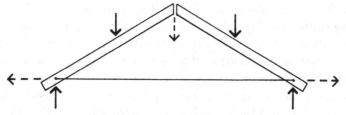

Rafters supported only by walls
exert thrust outward at top of walls.
Tie across building is one solution.

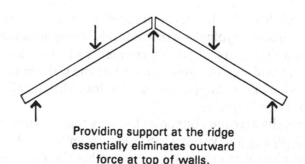

Providing support at the ridge
essentially eliminates outward
force at top of walls.

FIGURE 2.3 Stability of the gabled rafters.

2.2 THE FLOOR STRUCTURE

Development of the floor structure involves many of the same concerns that I discussed for the roof structure.

Plywood decking is typically at least $\frac{3}{4}$ in. thick, offering the added stiffness desired for a surface that will be walked on. However, the structural plywood deck is hardly ever the finished floor surface, so a lot depends on what constitutes the rest of the floor construction.

Floor joists can be selected from tables like those in Appendix B (see Table B.9). The usual live load for a residence is 40 psf; but 40 psf is a typical building code minimum requirement and a really tight design will probably guarantee a relatively bouncy floor. If you want solid-feeling floors, don't squeeze the design to the code minimum requirement.

Framed floors are usually separating elements—that is, you need to accommodate both the floor development above and a ceiling development below. Perhaps there are acoustic concerns or fire separation needs. In Building One, for example, the bedrooms over the garage may need a fire-rated floor construction. And the bedroom over the basement rumpus room may want some acoustic separation.

Actually, in this example most of the floors are not framed but are concrete paving slabs (usually called *slabs on grade*). For a residence, a slab on grade is usually 4 in. thick with minimal wire reinforcement for shrinkage stresses. This assumes, of course, that the paving support materials can be developed with a minimum of future settlement.

Supports for the floor structure, as for the roof, consist mostly of the building walls. Consideration for a special beam at the garage is discussed in Section 2.4.

2.3   THE WALLS

For a building of this size, the stud-framed wall with 2 × 4 studs at a maximum of 16 in. centers is typically a redundantly strong structure. In order to form wall ends and corners and the openings for windows and doors, extra studs are used, providing even more strength to the wall frames. This frame is more than adequate to support roof and floor structures and to frame shear walls for lateral bracing.

The hollow walls provide convenient spaces (called *interstitial space*) to conceal wiring, piping, and various recessed elements of the lighting, electrical power, and plumbing systems. Be careful that the installation of these elements does not result in excessive cutting away of the wall structure: the finished frame with service elements installed should be inspected prior to closing up the walls with surfacing materials.

To support large beams, it is possible to incorporate a larger timber section (4 × 4, 4 × 6, and so on) into the wall without revealing its existence. It is also possible simply to use multiple studs grouped together as a column or multiple joists as a beam.

In cold climates you may need to provide more insulation in exterior walls than can be accommodated by the 3.5-in.-wide space in the 2 × 4 stud wall. You can do so by using larger studs—2 × 6 or even 2 × 8—but also by adding wood strips to the face of the 2 × 4 studs.

Not much structural design is typically necessary for the walls in this size building. I demonstrate some issues that can occur in larger buildings in Parts Two and Three. A special concern is shear wall actions for lateral loads, which I discuss briefly in Sections 2.5 and 2.6.

2.4   SPECIAL ELEMENTS

Although most of the building construction for the light wood frame can be determined with little structural computation, some special structural elements are usually required for any building. For Building One, a few such items are the following:

The beam under the front wall of the upper level, which must be used if the garage is to be column-free.

The roof beam at the point between the living room and the front hall.

The posts that support the living room beam.

The beam over the wide door opening at the garage.

In the following four sections, I present some things to consider when designing these elements.

Note: References to help the design of these structural elements abound. The most comprehensive source currently availabe is the *Timber Construction Manual* (Ref. 5), although many other texts and handbooks are available. In this book, you can find examples of computations for wood elements for the buildings in Parts Two and Three.

Beam at the Garage Ceiling

This beam is rather long for such a small building. The load is not exceptionally large but does include the weight of the wall above. You can use a large timber section, but, for reasons of dimensional stability (keeping its shape after construction), you should consider some other options.

A steel beam, for example, offers long-term stability. It is a good structural choice but presents some difficult details of support in the wood frame structure unless steel columns are incorporated into the wood frame.

A glued laminated beam offers better dimensional stability (less sag, no splitting, and so on) and considerably greater strength than a solid sawn timber of the same size. Buying only one such beam is not very cost effective, but you may find other opportunities to use others in the building—perhaps the beams at the garage door and the living room roof.

Or you can use solid lumber and plywood to produce a special beam with a box cross section. The solid lumber serves as the equivalent flanges of an I-shaped beam, while the plywood serves as the beam web. In Building One, there is enough depth for such a beam if you use the vertical distance between the window sill and the garage ceiling.

Beam at the Living Room Roof

This beam cuts down the span of the rafters. Because of the sloped ceiling, there is considerable headroom, so having the beam extend below the rafters is not a problem. If the beam is dropped below the rafters, the rafters may be continuous over the beam if the required lengths of rafters can be obtained.

This beam can be a solid sawn section, a glued laminated section, or a steel shape. A steel beam would probably be boxed in to expose only its profiled shape. If the exposed framing option is used (see Figure 1.4), a wood beam would be used.

Post for the Living Room Beam

The living room beam may be supported at two points: at the kitchen wall and also somewhere within the open space, between the living room and entry area.

It is unlikely that a post is required at the corner of the kitchen, however, as at least three studs will be used to form this corner. And, at the corner, the wall provides its own bracing in both directions, so lateral stability is not a concern.

The freestanding post in the living room is essentially unbraced for its full height. This height is at about the limit for a 3.5-in.-wide member (limit of h/t = 50), so a 6 × 6 may be required, even though it will undoubtedly be redundantly strong. (see Figure B.1).

Where the beam ends rest on the exterior walls, adequate support can probably be provided by providing doubled studs, although they should be indicated as required on the construction drawings.

Beam at the Garage Door

The typical double garage door requires about a 17 ft span beam for the clear opening. This length is generally beyond the capacity of 2 in. dimension lumber (typically 2 × 12 maximum), so a real beam is required. You can use a timber, steel, glued laminated, or plywood box section.

Another option is to form a truss. In this case, you can use the whole gabled-end to produce a triangular profile truss. However, there may also be sufficient height to form a flat truss, as the clear opening is typically not very high.

Another option—a little antique, but still feasible—is to use a flitched beam, which is a combination of lumber elements and a steel plate, bolted together to work in tandem (see Detail J in Figure 1.3). While the steel plate increases the strength of this section, its real purpose is to reduce long-term deflection, which is likely to occur with solid sawn lumber.

2.5 DESIGN FOR WIND

The light wood frame is ordinarily constituted as a box-like structure to resist wind loads. Wall, roof, and framed floor structures with their structural sheathing produce structural diaphragms that can usually be used to develop a lateral-load resisting system.

Many basic building code requirements for the light wood frame and for attachment with nails or other fasteners provide for a notable minimum capacity of resistance to wind. However, the overall nature of the system and the interaction of its distinct parts should be investigated for any building. I describe this process in some detail for the wood frame buildings in Parts Two and Three, so I will not present it for this building. The methods of investigation demonstrated there can be applied to this light wood frame structure as well.

Because of their low weight, wood frame buildings often need anchorage for both horizontal and uplift forces due to wind. For example, special hold-down devices can keep the entire roof structure in place on its supporting walls during severe wind storms. Similarly, steel bolts anchor the building to the building foundations. And, of course, the foundations must be heavy enough, or buried sufficiently in the ground, to constitute an adequate anchorage base.

2.6 DESIGN FOR EARTHQUAKES

Basic building code requirements for wood construction usually provide for a minimum capacity of resistance to earthquake forces. Conventional construction in areas of high seismic risk involves enhancing the light wood frame to provide better lateral resistance. The form of investigation for earthquake forces, using criteria from the *Uniform Building code* (Ref. 1), is demonstrated for other building examples in this book.

The typical box-like construction for the light wood frame has two distinct advantages for earthquake resistance. First, the interactive diaphragms in the box system and the typically tough nature of wood-framed diaphragms—especially when surfaced with structural plywood—offer considerable resistance. Second, the wood's low dead weight (compared to masonry or concrete) directly affects the lateral seismic forces. With very little additional effort, you can use these advantages to produce good earthquake resistance for ordinary wood structures.

3

CONSTRUCTION DRAWINGS

The structural designer ordinarily documents the design in three ways: structural computations, written specifications, and structural working (or construction) drawings.

The computations must be legible, as others may need to review them. It is essential to use standard formats, terminology, and notation, and to cite references for any data or formulas used. The end products of the computations should be items that mirror information provided on the drawings or in the specifications. For example, the required size of a beam may be noted on a drawing. And the material to be used and any standards used for its production or installation should be noted in the specifications.

Specifications are written in fairly legal form; they may need to stand up to cross examination in court. In general, any items of fact that can be covered in the specifications should be done there and not on the drawings, mainly to avoid possible conflict. Whenever possible, the drawings should refer to the specifications for exact data.

The principal purpose of the drawings is to explain the layout and details of the construction. The drawings may be pictorial, that is, show parts of the construction in fairly realistic form. However, the drawings also may use symbols. But it is essential to use the existing standard forms so that the drawings do not confuse anyone who needs to use them. Construction drawings are no place for original artistic style.

For large buildings, a set of drawings called the *structural drawings* is produced. These drawings generally do not show the entire construction, but rather views of the stripped-down structure. Structural drawings are intended for use by the persons who fabricate and install the structure.

To fully understand the structural drawings, one should also see the complete architectural drawings—that is, detailed drawings that show the complete construction. Figures 1.2 through 1.5 are of this nature, although even those do not show everything—such as wiring and piping. Figure 1.6 is a structural drawing; only the basic structure is shown.

Structural drawings typically consist mostly of framing plans and cross sectional details. This chapter presents some examples, relating to Building One.

3.1 STRUCTURAL PLANS

Figure 2.1 is a plan used to illustrate the layout of the structural framing for the roof and upper-level floor. This style of plan is common, indicating individual beams or joists as single lines. The locations of the lines typically indicate the position of the beam centerline in the plan view (looking downward). If a beam is continuous, its line extends through the support. If a beam is discontinuous, the line stops just short of the support.

The objective of the *framing plan* is to indicate the horizontal location of each framing member. What is not shown are the profile (slope) of the rafters and how any connections are achieved. Details or other drawings must be provided for this information.

The plan should also indicate supports for the framing, including individual columns and any bearing walls. In Figure 2.1, all the walls beneath the framing are shown as dashed lines, although not all are actually supporting walls for the framing. Note: These drawings are somewhat abbreviated and used primarily to illustrate the form of the structure. For actual construction, it is necessary to study the work carefully and be sure that all critical information is given on drawings or in the specifications.

The *foundation plan* is usually drawn as a horizontal section cut above the basement floor, or a slab on grade floor, showing buried elements in dotted line. This type of plan is shown in Figure 3.1, which indicates the lower level of Building One. Such a plan may be used to indicate the floor structure as well as other subgrade construction. Figure 3.1 also shows a number of details that may be used in Building One.

3.2 STRUCTURAL DETAILS

Structural details give information that cannot be adequately provided in the general construction (architectural) drawings, on the structural framing plans, or in the specifications. Notation is used to identify items in the drawing and to give necessary instructions to the builder.

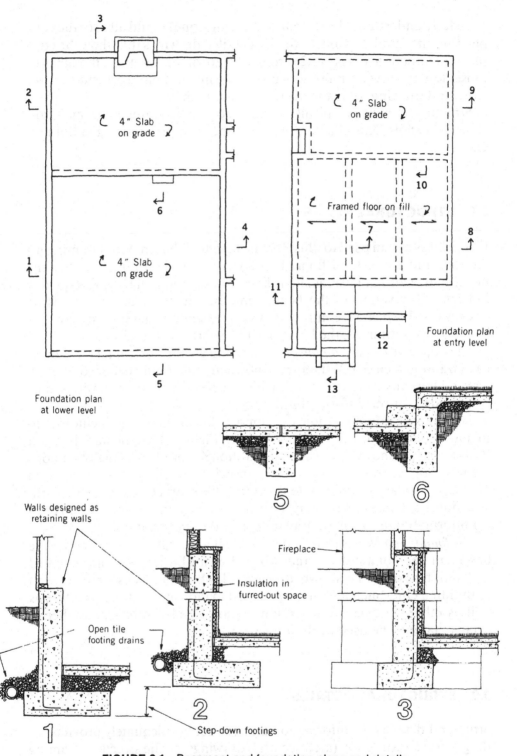

FIGURE 3.1 Basement and foundation plans and details.

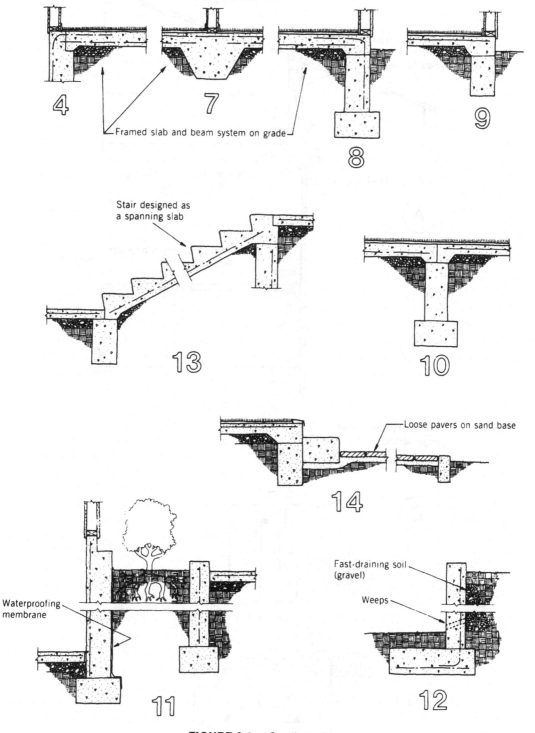

FIGURE 3.1 (*Continued*)

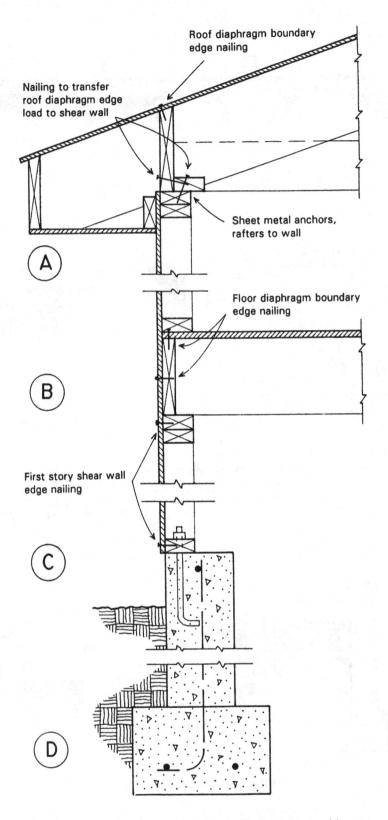

FIGURE 3.2 Typical structural details for the light wood frame.

Figure 3.2 includes details that might be used for the light wood frame structure. Reference should be made to corresponding details in Figures 1.2 through 1.5, where the full construction is shown.

Again, my purpose here is to explain the structure, not to propose models for professional contract documents.

PART II

BUILDING TWO

4

GENERAL CONSIDERATIONS
FOR BUILDING TWO

4.1 THE BUILDING

Building Two is a small, one-story building with a flat roof and a parapet wall of uniform height on all sides (see Figure 4.1). This is a modest-size building; one of a general form commonly used for a variety of commercial purposes. The permanent construction is often limited as much as possible to the parts of the building shell: the roof, exterior walls, and slab-on-grade floor. If generally structure-free, the interior can be endlessly manipulated by future occupants.

The building is generally symmetrical about a central corridor; one long side serves as the front (street-facing), the opposite long side as the rear. A short cantilevered canopy is attached to the front side, and the roof surface is sloped to the rear. For draining the roof surface, you most likely need scuppers through the rear parapet and vertical leaders on the rear wall.

As shown on Figures 4.2 and 4.3, the structure is a light wood frame with structural plywood siding on the rear and ends and brick veneer on the front. The floor is a concrete slab on grade. The roof structure consists of clear-span trusses (see Figure 4.4).

Use of the clear-span roof structure permits interior walls to be placed at will; remodeling is easy. (This form of structure is common for commercial buildings built as speculative rental properties.) The prefabricated trusses can have a flat bottom chord but also a sloping top chord to accommodate the roof drainage. The simplest ceiling construction consists of framing applied directly to the bottom chords of the trusses. In the void space between the roof deck and the ceiling, you can place ducts, piping, and wiring through the spaces between the truss diagonal web members.

To provide lateral bracing for wind or seismic effects, use the horizontal plane of the roof deck as a diaphragm, with vertical bracing from the plywood siding acting to form shear walls. Some detailing of the roof-to-wall joints, the framing of the roof deck and walls, and the foundations for the walls must respond to these lateral bracing functions.

Heating, cooling, and air conditioning is achieved with an all-air system, with ducts and ceiling registers in the void space between the roof and ceiling. HVAC equipment may go in the equipment room (see Figure 4.1) or on the rooftop, which is common in mild climates.

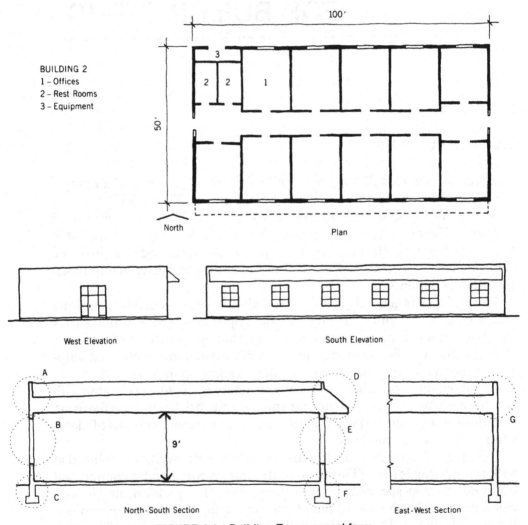

FIGURE 4.1 Building Two: general form.

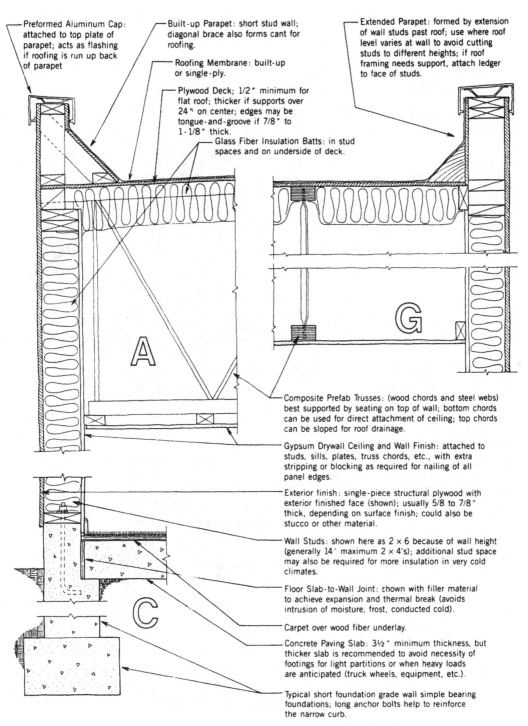

- Preformed Aluminum Cap: attached to top plate of parapet; acts as flashing if roofing is run up back of parapet

- Built-up Parapet: short stud wall; diagonal brace also forms cant for roofing.

- Roofing Membrane: built-up or single-ply.

- Plywood Deck; 1/2" minimum for flat roof; thicker if supports over 24" on center; edges may be tongue-and-groove if 7/8" to 1-1/8" thick.

- Glass Fiber Insulation Batts: in stud spaces and on underside of deck.

- Extended Parapet: formed by extension of wall studs past roof; use where roof level varies at wall to avoid cutting studs to different heights; if roof framing needs support, attach ledger to face of studs.

A

G

C

- Composite Prefab Trusses: (wood chords and steel webs) best supported by seating on top of wall; bottom chords can be used for direct attachment of ceiling; top chords can be sloped for roof drainage.

- Gypsum Drywall Ceiling and Wall Finish: attached to studs, sills, plates, truss chords, etc., with extra stripping or blocking as required for nailing of all panel edges.

- Exterior finish: single-piece structural plywood with exterior finished face (shown); usually 5/8 to 7/8" thick, depending on surface finish; could also be stucco or other material.

- Wall Studs: shown here as 2 × 6 because of wall height (generally 14' maximum 2 × 4's); additional stud space may also be required for more insulation in very cold climates.

- Floor Slab-to-Wall Joint: shown with filler material to achieve expansion and thermal break (avoids intrusion of moisture, frost, conducted cold).

- Carpet over wood fiber underlay.

- Concrete Paving Slab: 3½" minimum thickness, but thicker slab is recommended to avoid necessity of footings for light partitions or when heavy loads are anticipated (truck wheels, equipment, etc.).

- Typical short foundation grade wall simple bearing foundations; long anchor bolts help to reinforce the narrow curb.

FIGURE 4.2 Building Two: construction details.

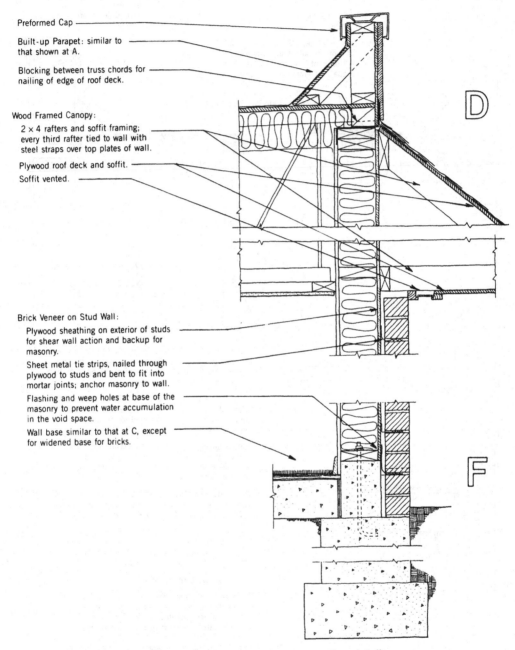

Preformed Cap

Built-up Parapet: similar to that shown at A.

Blocking between truss chords for nailing of edge of roof deck.

Wood Framed Canopy:

2 x 4 rafters and soffit framing; every third rafter tied to wall with steel straps over top plates of wall.

Plywood roof deck and soffit.

Soffit vented.

Brick Veneer on Stud Wall:

Plywood sheathing on exterior of studs for shear wall action and backup for masonry.

Sheet metal tie strips, nailed through plywood to studs and bent to fit into mortar joints; anchor masonry to wall.

Flashing and weep holes at base of the masonry to prevent water accumulation in the void space.

Wall base similar to that at C, except for widened base for bricks.

D

F

FIGURE 4.3 Building Two: construction details.

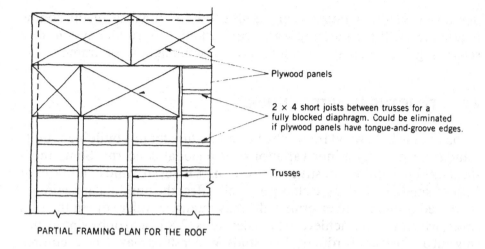

Plywood panels

2 × 4 short joists between trusses for a fully blocked diaphragm. Could be eliminated if plywood panels have tongue-and-groove edges.

Trusses

PARTIAL FRAMING PLAN FOR THE ROOF

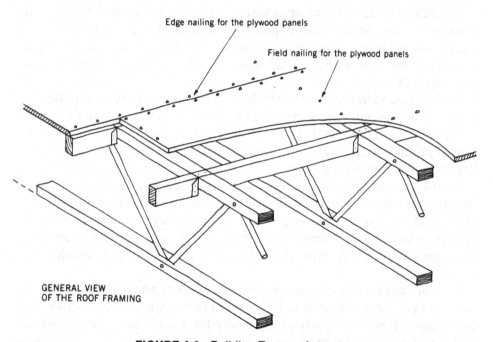

Edge nailing for the plywood panels

Field nailing for the plywood panels

GENERAL VIEW OF THE ROOF FRAMING

FIGURE 4.4 Building Two: roof structure.

Plumbing, consisting mostly of service for the rest rooms, is of minor concern. But you may also add a piped, underground drain system for the roof drains and a fire sprinkler system in the ceiling. In colder climates you may want to add a perimeter baseboard heating system to augment the air system in the ceiling.

Because this is a small building in total floor area, most building codes permit use of the light wood frame. However, the construction may have to develop a one-hour rating, depending on the occupancy or on zoning requirements. If the building was larger, adding a sprinkler system or a separating fire wall may still allow you to use the light wood frame.

4.2 STRUCTURAL ALTERNATIVES

If you use the light wood frame, the basic structure for this building is quite standard, with only minor variations for regional concerns. Some modifications are required for structural responses to severe wind conditions or high seismic risk, but these changes rarely affect the building appearance. Increased insulation in exterior walls may require a wider void in the stud space, which you can achieve with wider studs (2 × 6 or 8) or by simply furring out on the face of ordinary 2 × 4 studs. Wider studs may also be required because of the wall height, as shown here with 2 × 6 studs.

You can vary the exterior appearance endlessly by using different wall finish materials, window forms and materials, roof edge treatments, and detail features (such as the canopy). A canopy on all four sides, for example, can create the appearance of a building with a shingled or tiled, fast-draining roof form.

The light wood frame is probably the most architecturally manipulatable form of construction. You can use other forms of construction, of course, such as tilt-up concrete walls, exposed steel frame, structural masonry walls, or heavy timber frame; all will produce some specific forms and details that affect the building appearance.

As with any building, the choice of window materials and details, depth of foundations below grade, need for insulation of the floor slab at the building edges, and other items may require special consideration for extreme climate conditions. And exterior finish materials may be a regional concern, although just about any type of appearance can be achieved anywhere by some means.

Certain interior features become relatively fixed and are not subject to easy change: the building foundations and major structure, windows and doors, and plumbing—particularly sewer lines built into the pavement. Changes are easier if most services are installed in walls or in the space above the ceiling. This placement allows for major reworking of wiring, ducts, and various items for the building power, lighting, communication, and HVAC systems. When developing construction details, keep in mind both permanent elements and speculatively demountable elements.

If you affix the ceiling directly to the bottom chords of the trusses, the open webs of the trusses allow major elements, such as ducts, to run perpendicular to the trusses. A deeper-than-average truss may be used to place the ceiling at a desired level or to provide a maximum of space between truss web diagonals.

Of course, if a permanent interior wall system is planned, or if structure-free interior space in general is not significant, it well may be possible to develop a roof structure with some interior supports, as illustrated in Chapter 5. A clear-spanning wood truss system is also illustrated in Chapter 5. In Chapters 6 and 7 clear-spaning systems are described using steel elements.

5

DESIGN OF THE WOOD STRUCTURE

In this chapter I describe the design of an all-wood structure for Building Two. Although some of the drawings in Chapter 4 show a clear-spanning truss roof structure, I show in this chapter how to support the roof with interior supports.

I also illustrate how to use composite wood and steel joists (wood chords and steel web); I detail the design of a clear-spanning system with steel open web joists in Chapter 6. These basic systems are similar in their support, lateral bracing, support of decks and ceilings, and so on.

5.1 THE ROOF STRUCTURE

With the construction as shown in Figure 4.2, the roof loads are as follow:

Three-ply felt and gravel roofing	5.5 psf
Fiber glass insulation batts	0.5
$\frac{1}{2}$-in.-thick plywood roof deck	1.5
Rafters and blocking (estimate)	2.0
Ceiling joists	1.0
$\frac{1}{2}$ in. drywall ceiling	2.5
Ducts, lights, etc.	3.0
Total roof dead load	16.0 psf

Assuming a partitioning of the interior as shown in Figure 5.1*a*, various possibilities exist for the development of the spanning roof and ceiling framing systems and their supports.

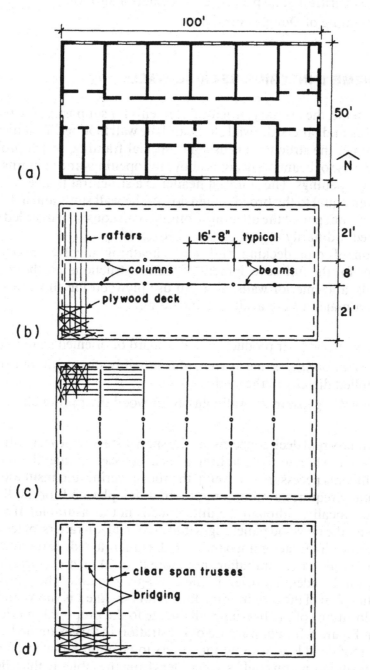

FIGURE 5.1 Alternative roof structures: (*a*) Partitioning system with a double-loaded, central corridor. (*b*) Post, beam, and rafter system with long-direction beams. (*c*) Post, beam, and rafter system with short-direction beams. (*d*) Clear-spanning truss system.

You can assume the following data for design:

Roof live load = 20 psf (reducible).
Wind load critical at 20 psf on vertical exterior surface.
Wood framing of Douglas Fir–Larch.

5.2   SCHEME 1: INTERIOR BEARING WALLS

For this scheme, the primary structural elements for supporting the roof are the roof deck and rafters, the wall studs, and the wall footings. To achieve the wall openings, the structure needs some special framing, which probably involves the use of beams as headers above the openings and columns at the sides of the openings. The simplest header is a structural lumber member with a width equal to the broad dimension of the wall studs and a depth as required for the span of the opening. Columns can consist of doubled studs, a form used ordinarily at wall ends and edges of openings.

Selection of roof decking and wall sheathing involves many considerations for the full development of the constructions for the roof, exterior walls, and interior walls. You can use plywood for all surfaces, but many other materials are available, including:

Roof deck: wood fiber products of flake board or oriented strand board.
Exterior wall surface: same as roof, plus possibly stucco without sheathing, applied directly to the studs.
Interior walls: gypsum drywall, plaster, or wood fiber products.

The primary roof deck concerns are the span distance between rafters, the resistance to the horizontal diaphragm shear stresses for lateral forces, and the attachments necessary to develop the roof covering and insulation. The latter usually requires a minimum of $\frac{1}{2}$-in.-thick plywood for the flat roof (now $\frac{15}{32}$ technically, although the difference is not measurable). If you use blocking for the plywood panel edges that do not fall on rafters, place the 4 ft × 8 ft sheets with the face grain parallel to the rafters to reduce the amount of blocking. In this case, with rafters not more than 24 in. on center, the $\frac{1}{2}$-in. plywood will be adequate, even in the cross-grain span of the panels.

A partial plan of the roof framing layout for Scheme 1 is shown in Figure 5.2. The locations of the bearing walls relate to the proposed interior plan shown in Figure 5.1a. For purpose of illustration in this example I assume the total roof dead load without the rafters to be approximately 23 psf.

For this situation, you could select rafters from the tables in the *UBC*. The problem is finding the proper table, with the correct combination of the design live load (20 psf) and the actual dead load (23 psf). Note that neither

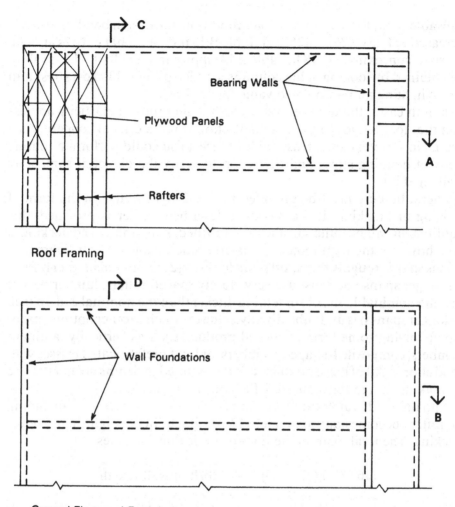

Roof Framing

Ground Floor and Foundations

FIGURE 5.2 Structural plans for system with interior bearing walls.

Table B.13 nor B.14 quite matches up. However, you can make a good guess by comparing the answers from both tables:

Table B.13 has a dead load 3 psf lower than 23 psf, so the answer produced is on the skimpy edge.

Table B.14 has a total load (DL + LL) 7 psf larger than 43 psf and a live load 50% greater. Since deflection under live load is often the limiting condition for maximum spans, the answer from this table is quite conservative.

For the No. 2 Douglas Fir–Larch rafters, the allowable bending stress for repetitive stress members is 1006 psi, and the modulus of elasticity is 1,600,000 psi (see Appendix B, Table B.4). With the usual 25% increase in the allowable stress for the roof live load condition, the design bending stress is increased to 1.25 (1006) = 1257 psi. Table B.13 indicates that 2 × 12s at 24 in. can span approximately 18 ft; Table B.14, approximately 16 ft. These values are obtained by looking at the columns for 1200 psi and 1300 psi stress and approximating an interpolated value for 1257 psi.

In both cases, the required values for the modulus of elasticity are less than that for the No. 2 rafters, so deflection is not a critical concern. As a result, No. 2 rafters are a reasonable choice. (You could perform a routine check of bending stress and deflection to verify the choice if somebody challenges it.)

Lateral bracing must be provided for the 2 × 12 rafters in the form of bridging or blocking. If the blocking described earlier is provided, this requirement will be quite satisfactorily fulfilled. If an unblocked decking is used, however, the requirement for lateral bracing must be considered.

This span is roughly the cutoff point for using 2 × nominal lumber rafters. For longer spans you must use very closely spaced rafters, lumber deeper than the nominal 12 in., or something thicker than the nominal 2 in. In fact, for longer spans it is undoubtedly advisable to switch to other options, probably involving some form of special product, such as vertically laminatd members, composite I-shaped members, or light prefabricated trusses with wood chords. All of these products can be obtained in depths greater than 12 in., making design for reduced deflections more feasible.

With the 2 × 12 rafters at 16-in. centers, the total roof dead load including the rafters becomes approximately 27 psf, including a slight increase for blocking. The total load on the interior walls thus becomes

$$(16.67 \text{ ft}) (27 + 20) = 784 \text{ lb/ft of wall length}$$

This load can easily be borne by 2 × 4 studs at 16-in centers if the wall height is not too great. These walls, however, are major dividing walls between tenant spaces, and their general sturdiness and/or desired acoustic separation function may indicate some different choices. The minimum typical partition with 2 × 4 studs and gypsum drywall on both sides is really quite flimsy.

Adding the roof load to the weight of the wall produces a load of less than 2000 lb/ft at the bottom of the wall. Although a wall footing should be provided for these bearing walls, its width will be quite minimal in this case, even with the very low allowable soil pressure of 2000 psf.

The exterior walls on the long sides of the building (parallel to the rafters) carry mostly only their own dead weight and that of the roof's edge. The walls on the short sides, however, carry these loads plus a half-span of the rafters. In addition, the exterior stud walls must resist the direct wind pressure or lateral seismic force due to their own weights. The design of these studs therefore must take into account the combined axial compression and bending condition (see Section 5.4).

5.3 SCHEME 2: BEAM FRAMING WITH INTERIOR COLUMNS

Although interior walls may be available for use as supports, a more desirable situation in many commercial interiors is the use of a clear-spanning structure or one with widely spaced columns to permit future rearrangement of interior partitioning. The roof framing system shown in Figure 5.1*b* includes two rows of interior columns placed at the location of the corridor walls. If the interior plan shown in Figure 5.1*a* is used, these columns are incorporated in the wall construction and out of view.

Figure 5.1*c* shows a second possibility for the roof framing using the same column layout as in Figure 5.1*b*. Both framing schemes work; ducts, lighting, wiring, roof drains, and fire spinklers may influence your choice. I arbitrarily selected the scheme shown in Figure 5.1*b* to illustrate the design process for this structure.

Installation of a membrane-type roofing ordinarily requires at least a $\frac{1}{2}$-in.-thick roof deck. Such a deck is capable of up to 32 in. spans in a direction parallel to the face ply grain (the long dimension of ordinary 4 ft × 8 ft panels). If rafters are not over 24 in. on center—as they are likely to be for the schemes shown in Figures 5.1*b* and *c*—the panels may be placed so that the plywood span is across the face grain; such an arrangement reduces the blocking required at edges not falling on the rafter.

For the rafters I assume No. 2 grade, an ordinary minimum structural usage. The allowable bending stress for repetitive use is 1006 psi, and the modulus of elasticity is 1,600,000 psi (see Table B.4). Since the loading falls approximately within the criteria of Table B.13, that table can serve as a reference. From it, you can see that the rafters should be either 2 × 10s at 12 in. centers or 2 × 12s at 16 in. centers. You can verify the adequacy of either choice by using simple computations. For stress investigation it is likely that a 15% or 25% increase of the allowable stress would be permitted.

If the ceiling joists also span the full 21 ft, Table B.10 using either 2 × 8s at 16 in. centers or 2 × 10s placed at 19.2-in. centers. Spacing of ceiling joists must, of course, relate to the surfacing material. It is also possible to either suspend the ceiling joists from the rafters or use the interior partitions for support. The dead load tabulation made previously assumes the use of light ceiling framing—most likely, 2 × 4 or 2 × 3 members—suspended from the rafters.

The beams shown in Figure 5.1*b* are continuous through two spans, with a total length of 33 ft 4 in. and a clear span of 16 ft 8 in. For the two-span beam, the maximum bending moment is the same as that for a simple span; the principal advantage is a reduction of deflection. The total load area for one span is

$$16.7 \times \frac{21 + 8}{2} = 242 \text{ ft}^2$$

As indicated in Table A.2, this total permits the use of a live load of 16 psf. Thus the unit of the uniformly distributed load on the beam is

$$(16 + 16) \times \left(\frac{21 + 8}{2} \right) = 464 \text{ lb/ft}$$

You should add a bit for the beam weight and design for a load of 480 lb/ft. The maximum moment thus is

$$M = \frac{wL^2}{8} = \frac{480(16.67)^2}{8} = 16{,}673 \text{ ft-lb}$$

The allowable bending stress depends on the beam size and the load duration assumed. If you assume a 15% increase for load duration the allowable bending stress is (see Table B.4)

For a 4× member: $F_b = 1.15(1000) = 1150$ psi
For a 5× member or larger: $F_b = 1.15(1300) = 1495$ psi

Then

$$S = \frac{M}{F_b} = \frac{16{,}673 \times 12}{1150} = 174 \text{ in.}^3$$

so a 4 × 16 is not quite adequate ($S = 135.7$ in.^{3}). But

$$S = \frac{16{,}673 \times 12}{1495} = 134 \text{ in.}^3$$

so either a 6 × 14 ($S = 167$ in.^{3}) or an 8 × 12 ($S = 165$ in.^{3}) is adequate.

Although the 4 × 16 offers the least cross section area and ostensibly is cheapest, several things may affect the beam selection. For example, the beam can be formed as a built-up member from a number of 2× elements, which may prove vital where good-quality heavy timber sections are not easily obtained. Or the beam may consist of a glued laminated or rolled steel section, reducing beam depth and gaining a better long-term deflection response, which is more critical for beams of longer span (for example, the beams in the framing scheme in Figure 5.1c).

The beam should also be investigated for shear and deflection. Note that the maximum shear in the two-span beam is slightly larger than the simple beam shear of $wL^2/2$. This may be a reason for *not* using the two-span condition in wood beams, where shear is often a critical concern for solid-sawn sections. In this example, the span is quite short, so deflection is not a critical concern, especially if the two-span condition is used. Shear is also not critical if the stress increase for load duration is made and the load on the ends of the spans is eliminated.

Assume that a minimum roof slope of $\frac{1}{4}$ in. per ft is required to drain the roof surface. If the draining occurs at the outside walls, the slope length is only one half of the building width. The total elevation change is approximately $\frac{1}{4} \times 25 = 6.25$ in. from the center to the long side of the building. You can achieve this slope in various ways, including the tilting of the rafters.

Figure 5.3 shows some possibilities for the buildings center. Note that the rafters are kept flat and the roof slope is achieved by attaching cut 2× elements on top of the rafters and using short profiled rafters at the corridor.

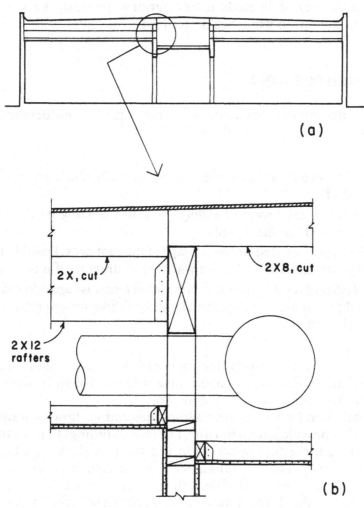

(a)

2X, cut

2X8, cut

2X12
rafters

(b)

FIGURE 5.3 Details for structure with interior posts.

Ceiling joists for the corridor are supported directly by the corridor walls. Other ceiling joists are supported at their ends by the walls, although long-span ceiling joists could be suspended in their spans by hangers from the rafters. You can develop ceiling construction together with the roof structure or with the interior partitioning, depending on the occupants' needs.

The typical interior column supports a load approximately equal to the entire load on one beam or

$$P = 480 \times 16.67 = 8000 \text{ lb}$$

This is quite a light load, but the column height requires larger than a $4 \times$ size (see Figure B.1). If a 6×6 is not objectionable, it is adequate in the lower-stress grades. However, consider using a steel pipe or tubular section; either can be accommodated in a partition wall with 2×4 studs.

Structural design of the studs in the exterior walls must take into account lateral bending due to wind. See Section 5.4.

5.4 DESIGN FOR WIND

Designing the building structure for wind requires consideration of the following:

1. Inward and outward pressure on exterior walls, causing bending of the wall studs.
2. *Total lateral force on the building.* Requires bracing by the roof diaphragm and the shear walls.
3. *Uplift on the roof.* Require anchorage of the roof deck. Possibly requires some extra anchorage for the whole roof structure if it is very light.
4. *Total effect of combined lateral (horizontal) force and uplift (vertical) force.* Possibly requires anchorage of the building for overturn or horizontal sliding.

Although using a relatively low grade of 2×6 studs at 24-in. center-to-center spacing is adequate, an alternative is to use a slightly higher grade, reduce spacing, and make 2×4 studs work.

The total lateral force on the building determines how to design those elements that provide lateral bracing. For most buildings that use this form of construction, the first option is to use the roof deck as a horizontal diaphragm and the exterior walls as shear walls. If the roof has many large openings or uses some form of deck that is not rated for diaphragm shear capacity, some other form of horizontal spanning structure may be necessary. If there are too many gaps between solid wall portions—or if the walls are not solid (they are glazed, for example)—some other vertical bracing may be required. This example shows a building form and type of construction that presents a classic case of the so-called *box system building*. The "box" uses a roof plane that constitutes a horizontal diaphragm and walls that comprise a shear wall system.

How much you compensate for uplift force on the roof varies with different building codes. The maximum consideration usually consists of an uplift equal to the horizontal design pressure. Because the uplift force exceeds the roof dead load by 4 psf in this example, the roof must be anchored to its supports to resist this effect. Using metal-framing anchors between the rafters and the beams or between the rafters and the stud walls should provide adequate anchorage.

Overturn is not a critical concern for Building Two. Overturn is more critical in buildings of extremely light dead weight or that are tall and narrow. In any event, the investigation is similar to when investigating for overturn of a single shear wall, except you must consider the uplift effect on the roof. If the total overturning moment is more than two-thirds of the available resisting moment due to building dead load, the anchorage of the whole building must be considered.

When considering the forces exerted on the principal bracing elements— that is, the roof diaphragm and the exterior shear walls—the building must be investigated for wind in the two directions: east–west and north–south. Consideration of the wall functions and determination of the forces exerted on the bracing system are illustrated in Figure 5.4. The amount of the portion of the pressure on the exterior walls that is applied as load to the edge of the roof diaphragm depends partly on the spanning nature of the exterior walls. Figure 5.4*a* shows two common cases: with studs cantilevering to form the parapet and with simple span studs and a separate parapet structure. With the construction as shown in Figure 4.2, the walls on the north and south sides typify Case 2 in Figure 5.4*a*. Thus the lateral wind load applied to the roof diaphragm in the north–south direction is

$$(20 \text{ psf}) \left(\frac{10.5}{2} \right) + (20 \text{ psf})(2.5) = 155 \text{ lb/ft}$$

In resisting this load, the roof functions as a spanning member supported by the shear walls at the east and west ends of the building. The investigation of this 100-ft-span simple beam with uniformly distributed loading is shown in Figure 5.5. The end reaction and maximum shear force is found as

$$(155) \left(\frac{100}{2} \right) = 7750 \text{ lb}$$

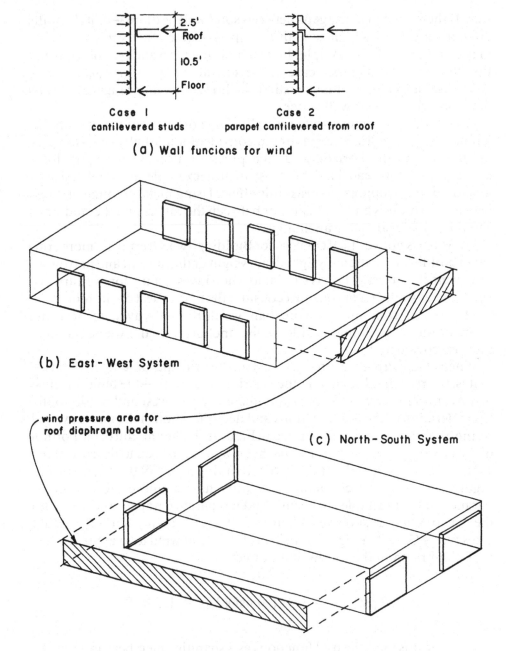

FIGURE 5.4 Development of wind loads and the wind-resisting structure.

which results in a maximum unit shear in the 50-ft-wide roof diaphragm of

$$v = \frac{\text{shear force}}{\text{roof width}} = \frac{7750}{50} = 155 \text{ lb/ft}$$

From Table B.8 (*UBC*, Table 23-I-J-1) you can select a number of possible choices for the roof deck. Variables include the class of the plywood, the panel thickness, the width of supporting rafters, the nail size and spacing, and the use (or omission) of blocking. When choosing the roof deck construction, you must not only satisfy the shear capacity requirement, but also consider the design for gravity loading, general construction details for the roof structure, and the problems of installing roofing and insulation. For the flat roof with ordinary tar and felt membrane roofing, it is usually necessary to provide a minimum of $\frac{1}{2}$ in. (now $\frac{15}{32}$ in.) thick plywood. If this requirement is accepted, a possible choice from Table B.8 is

> Structural II $\frac{15}{32}$ in. plywood with 2 × framing and 8d nails at 6 in. at all panel edges and a blocked diaphragm.

For these criteria Table B.8 yields a capacity of 270 lb/ft.

In this example the minimal construction is more than adequate for the required lateral force resistance. If the required capacity results in considerable nailing beyond the minimal, it is possible to graduate the nail spacings from the maximum required at the building ends to minimal nailing in the center portion of the roof.

The moment diagram shown in Figure 5.5 indicates a maximum value of 194 k-ft at the center of the span. This moment is used to determine the required chord force that must be developed in both tension and compression at the roof edges. With the construction as shown in Figure 4.2, the top plate of the stud wall is the most likely element to be used for this function. In this example the force is quite small and can be easily developed by the ordinary construction. The one problem that may require some special effort is the development of the member as a continuous tension element, since it is not possible to have a single 100-ft-long piece for the plate. You must splice the multipiece plates to provide a continuity of the tension force.

The end reaction force for the roof diaphragm, as shown in Figure 15.5, must be developed by the end shear walls. As shown in Figure 15.1*a*, there are two walls at each end, both 21 ft long. Thus the total shear force is developed by a total of 42 ft of shear wall. The unit shear in the walls is

$$v = \frac{\text{total shear force}}{\text{total wall length}} = \frac{7750}{42} = 185 \text{ lb/ft}$$

When choosing the wall construction, you must consider many variables. A common situation involves the use of a single facing of structural plywood on the exterior surface of the wall, which is considered as the resisting element for lateral force. Other elements of the wall construction thus are considered nonstructural in function. For this situation a possible choice from Table B.7 (*UBC*, Table 23-I-K-1) is

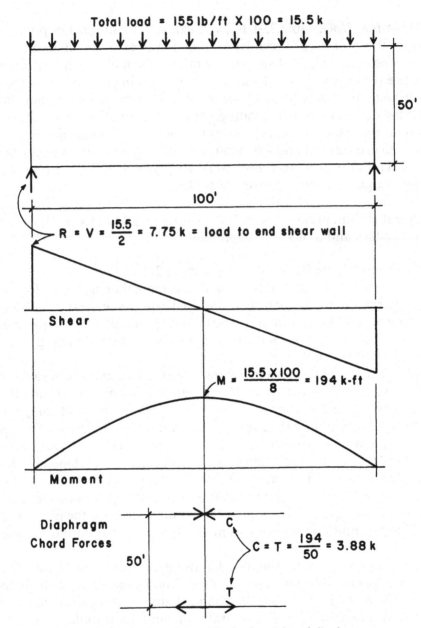

Total load = 155 lb/ft × 100 = 15.5 k

50'

100'

$R = V = \dfrac{15.5}{2} = 7.75\,k$ = load to end shear wall

Shear

$M = \dfrac{15.5 \times 100}{8} = 194\,k\text{-ft}$

Moment

Diaphragm
Chord Forces

50'

C

$C = T = \dfrac{194}{50} = 3.88\,k$

T

FIGURE 5.5   Functions of the horizontal roof diaphragm.

Structural II, $\frac{3}{8}$ in. thick plywood with 6d nails at 6 in. spacing at all panel edges.

Again, this is minimal construction. For situations that require much higher capacities of shear resistance, it may be necessary to use thicker than normal plywood and nailing with larger, closer spaced nails. Unfortunately, the nailing cannot be graduated because the unit shear is a constant value throughout the height of the shear wall.

Figure 5.6*a* shows the loading condition for investigating the overturning effect on the shear wall. The loading shown includes only the lateral force (applied at the level of the roof deck) and the dead loads of the wall itself and the portion of the roof carried by the wall. The overturning moment is the product of the lateral force and its distance above the wall base, and a factor of 1.5, which represents the usual minimum required safety factor for comparison with the dead load resisting moment (also called the *restoring moment*). If the minimum safety factor is not provided, an anchorage force (*T* in Figure 5.6*a*) must be added to supplement the dead load resistance. The usual computations for this investigation are as follow:

$$\text{overturning moment} = (3.875)(11)(1.5) = 64 \text{ k-ft}$$

$$\text{restoring moment} \quad = (3 + 6)(10.5) = 94.5 \text{ k-ft}$$

These computations indicate that no tiedown force is required.

Actually there is additional resistance at each end of the wall: at the corner this wall is most likely well attached to the wall on the north or south side of the building, providing additional dead load resistance. In addition, at the end of the wall near the corridor is a post that supports the beam (see the framing plan in Figure 5.2); the dead load portion of the beam reaction provides additional resistance.

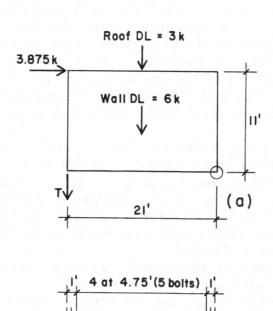

FIGURE 5.6 Actions of the shear wall.

Finally, the wall sill is bolted to the foundation, providing some hold-down resistance to toppling of the wall. At present, most codes do not permit use of a computed value for the sill bolt resistance to uplift because this function involves cross-grain bending of the sill.

The sill bolts are used for resistance to the horizontal sliding of the wall, however, and the bolting must satisfy this requirement. Code minimum bolting usually consists of $\frac{1}{2}$ in. diameter bolts at 1 ft from the wall ends and at a maximum of 6 ft on center for the remainder of the wall length. A layout for this minimal bolting is shown in Figure 5.6b. With a $2\times$ sill member and $\frac{1}{2}$ in. bolts in single shear, UBC tables yield a value of 470 lb for one bolt. With five bolts and an increase of one-third in the value for wind loading, the total sliding resistance of the minimum bolting is

$$(1.33)(470)(5) = 3125 \text{ lb}$$

As this resistance is a bit short of the necessary resistance, you must increase the number and/or the size of the bolts. It is probably more economical to increase the bolt size: the cost of setting bolts in concrete is not much related to bolt size. Thus a few large bolts are preferable to a lot of small bolts.

In some situations it may be necessary to consider the effects of the sliding and overturn forces on the wall foundations. In this example the forces are quite low and not likely to be critical. For buildings with shallow foundations, however, the overturning and sliding must be seriously considered, especially when they are high in magnitude in relation to gravity forces.

In designing for lateral force resistance, you should consider the necessary force transfers between the various elements of the lateral bracing system. The sill bolting of the shear wall is one example of such a transfer. Another critical transfer is that of force from the roof diaphragm to the shear wall. In this example the roof shear is vested in the roof plywood and the wall shear in the wall plywood. These two elements are not directly attached, so the details of the framing must be studied to determine how the transfer can be made.

With the construction as shown in Figure 4.1, a simple transfer can be made through the top plate of the wall to which both the roof and wall plywoods are directly attached. This is likely to occur normally for the roof plywood when this is the edge of the roof diaphragm (called the *boundary* in Table B.8). For the wall plywood, however, this location may not be a panel edge, and the necessary nailing for the required transfer should be indicated and specified on the wall section.

For other wall and roof constructions, the force transfer may not be as simple and direct as in Figure 4.2. In some cases there may not be any direct transfer through ordinary elements, thus requiring additional framing or connecting devices.

5.5 SCHEME 3: CLEAR-SPANNING ROOF TRUSSES

Figure 5.7 shows a partial framing plan and some details for a modified wood roof structure. This structure uses light, composite wood and steel trusses to achieve a clear span of the 50-ft-wide building, eliminating all interior structural supports. Such a structure offers flexibility of interior space arrangements. For commercial buildings whose space will be rented, this flexibility is a major goal.

The partial framing plan in Figure 5.7 shows the layout for this system, indicating truss spacing of the trusses. For practical purposes, the lateral bracing also functions to assure accurate spacing, vertical straightness, and longitudinal straightness of the flexible trusses. The trussed system is quite sturdy when properly erected, but the correct bracing of the system is essential.

These trusses undoubtedly can be the proprietary products of some manufacturer; such products are widely available. Refer to safe load tables prepared by the manufacturer when selecting the trusses. Be sure to consider the nominal depth of the trusses and the sizes of individual members—in particular, the top and bottom wood chords.

Truss spacing relates to not only truss load capacity but also the roof deck and ceiling construction. A 2 ft spacing permits use of ordinary plywood panels for the roof deck and gypsum drywall for the ceiling: both surfaces are nailed directly to the truss chords. The truss chords may be nonparallel to achieve the sloping roof and the flat ceiling. This is undoubtedly the simplest and most economical form of construction.

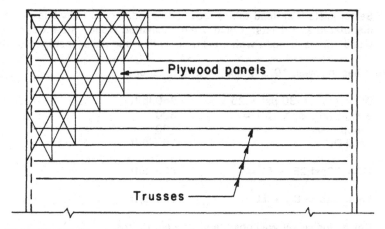

FIGURE 5.7 Partial framing plan for the clear-spanning truss system.

However, 2 ft spacing is a bit close for this span, so the scheme shown here uses a 32 in. spacing (still a module of the 96 in. standard plywood panel) with a $\frac{7}{8}$-in.-thick plywood deck. These panels are available with tongue-and-groove edges, which permit the development of an unblocked deck. Need for blocking, however, also relates to the development of the roof deck as a horizontal diaphragm.

The long-span trusses generate a significant vertical load for the stud walls on the building's long sides. The combination of this vertical loading plus the direct wind load on the wall surface requires a design for bending plus axial compression for the studs. The conditions for this and the other major elements of the structure are summarized in Figure 5.8. Example computations for these elements are developed in other parts of the book as indicated in Figure 5.8.

Example computation summary for Scheme 3, Building 2

1. Typical roof deck, plywood panels, 32" span, face grain perpendicular to the supports.

    Table B.5 (UBC Table 23-I-S-1), 40/20 panel, 7/8" thick,
    allowable span of 32" if unblocked (edges not supported)

2. Joist (truss), at 32" on center, 50 ft span (+ or -).

    LL at 20 psf: 20(32/12) = 53.3 lb/ft

    DL: roofing + insulation + deck + ceiling + equipment, say 25 psf
    25(32/12) = 66.7 lb/ft (not including joists)

    Total load = 53.3 + 67.7 = 121 lb/ft

    Select joists from load tables provided by industry or individual manufacturer of joists. Could use combination wood + steel truss with wood chords or open web steel truss with wood nailer fastened to top.

3. Front bearing wall, 10.5 ft high, with 5 ft canopy.

DL = roof at (30 psf X 25') =	750 lb/ft
wall at (20 psf X 15') =	300
canopy, say	100
Total DL =	1150 lb/ft
LL = 20psf(25 + 5) =	600 lb/ft
Total load = DL + LL =	1750 lb/ft

    Design individual wall posts and headers for the wall load.
    This is also the load on the footing.

FIGURE 5.8   Summary of computations for the truss roof structure.

5.6 FOUNDATIONS

For ordinary soil conditions, bearing foundations for Building Two are quite minimal. Exterior wall foundation construction depends on possible frost penetration and the location (depth) of good, undisturbed bearing materials.

In warm climates, when good bearing material is available close to the finished grade level, you may use a shallow grade beam that combines the functions of a foundation wall and footing (see Figure 5.9a). The depth of this beam depends on the minimum depth required below finished grade, and its width depends on the allowable soil bearing pressure.

In cold climates, or where good soil is some distance below finished grade, you may use a separately cast footing, with a foundation wall of variable height formed on top of the footing (see Figure 5.9b). This wall can also be achieved with concrete blocks (CMUs), as shown in Figure 6.1.

With closely spaced columns, supporting only light roof loads, column footings for the wood structure are quite small. Sizes may be chosen from Table B.22.

For this size building, foundations comprise concrete of low strength and a minimum of steel reinforcement. For even greater economy, some code-enforcing agencies don't require testing of the concrete if a design strength of

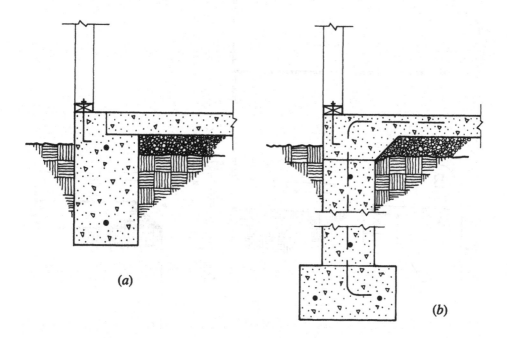

(a)

(b)

FIGURE 5.9 Options for exterior wall foundations.

only 2000 psi is used for the foundation elements. While this is below any desirable concrete quality, the economical design thus achieved would probably actually be done with concrete specified with a minimum of 3000 psi. If testing of concrete is not required, the savings will allow for a lot of conservative shortcuts in sophisticated engineering of the foundations.

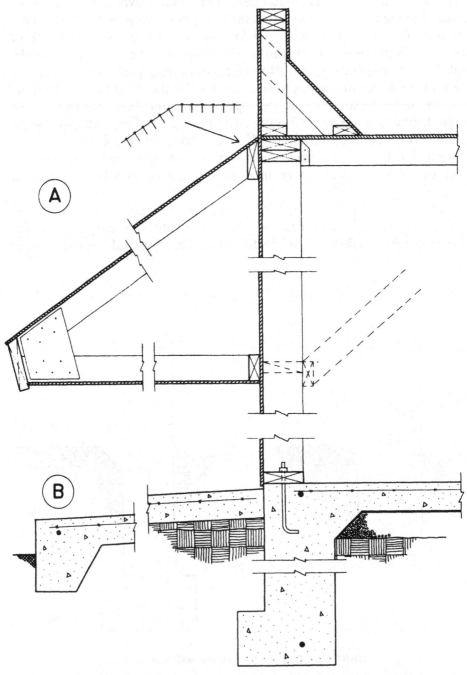

FIGURE 5.10 Construction details for the light wood frame structure.

5.7 CONSTRUCTION DETAILS

Compare the structural details in Figures 4.2 and 4.3 to Figures 5.10 and 5.11. Figures 4.2 and 4.3 show the condition with the clear-span trusses; Figures 5.10 and 5.11 show details with ordinary lumber framing for the roof.

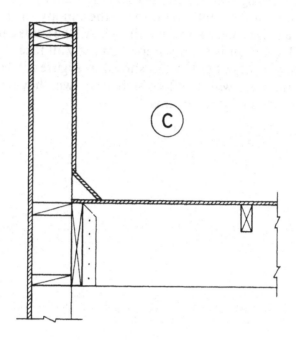

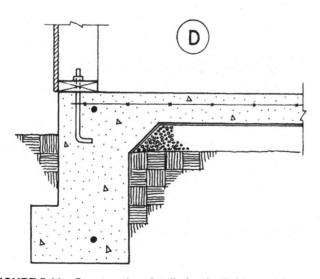

FIGURE 5.11 Construction details for the light wood frame structure.

Many variations are possible for the situation in Detail A in Figure 5.10. At issue is how the parapet and canopy are formed; both must cantilever from the basic roof/wall structure. The bottom of the canopy also creates an inward thrust on the wall studs, which might be resolved with struts, as shown in dashed line in Detail A.

Detail C in Figure 5.11 shows the condition at the end of the building, where the roof level changes along the length of the wall. Figure 5.11 shows the simplest solution: run the studs to the top of the canopy and support the end of the roof on a ledger placed on the wall face. All the studs can be cut to the same length. The ledger is simply sloped on the wall face.

Foundations may be of either type, as shown in Figure 5.9. However, if they are cast continuously with the floor slab as shown, they will be quite shallow and in the class of grade beams.

6

DESIGN OF THE STEEL
AND MASONRY STRUCTURE

Figure 6.1 shows an alternative construction for Building Two. The draw-ings indicate the use of structural masonry walls with concrete masonry units (CMUs) and a roof structure using open web steel joists and formed sheet steel deck. In this chapter, I discuss the designs for the clear-spanning roof truss structure and the CMU walls. Figure 6.2 presents the modified form of Building Two as used in this chapter.

6.1 THE STEEL ROOF STRUCTURE

Compare the partial plan of the roof framing system shown in Figure 6.3 to the plan for the wood clear-spanning system in Figure 5.7: notice that center-to-center truss spacing for the wood structure is typically quite close and uses a dimension that is an even division of the 8-ft-long plywood panels (12, 16, 24, 32, or 48 in.). With no such restraints here, the spacing between joists is based on other concerns, such as the following:

Span capacity of the formed sheet steel deck. For the minimum typical deck with corrugations or pleats 1.5 in. high, spans of 6 ft and more are easily achieved. (See Table B.15.)

Ceiling support system. Ceilings are usually hung from the bottom of the trusses.

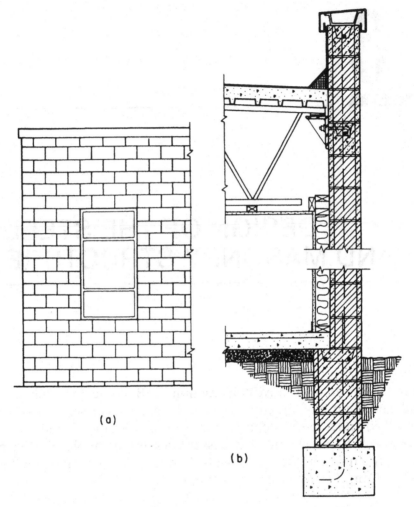

(a)

(b)

FIGURE 6.1   Details for the steel and masonry structure.

Joist (truss) depth and general weight. Both are a function of span and
spacing. If spacing is increased, the load is proportionately increased.
Some study may be made of alternative systems in this regard, using
more light joists closer together or fewer heavier joists farther apart.
(See Table B.16.)

Assuming a total dead load of 25 psf (roofing + insulation + ceiling + building equipment), and a live load of 20 psf, examine Table B.15 for possible
deck choices. Deck units are typically available in lengths of 30 ft or so, so the
likely span condition is "Three or More." As a result, viable choices are
NR20, IR20, or WR22 units.

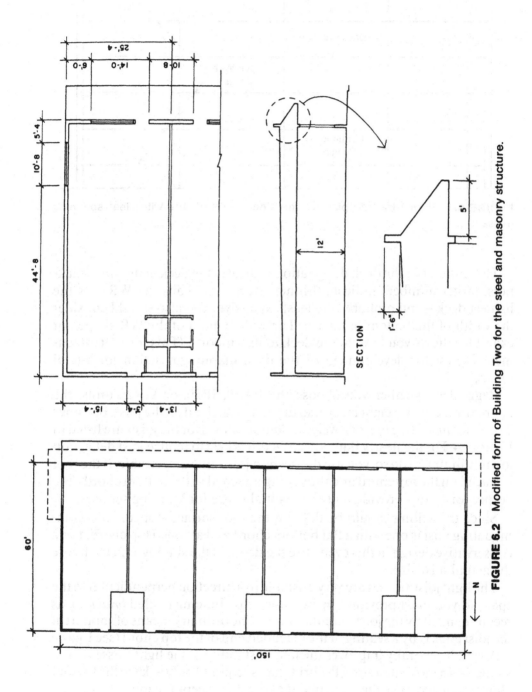

FIGURE 6.2 Modified form of Building Two for the steel and masonry structure.

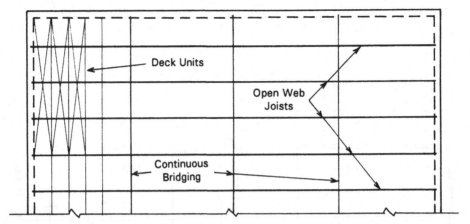

FIGURE 6.3 Partial framing plan for the steel roof structure with clear-spanning trusses.

The number in the deck designation indicates the deck gauge and thickness; larger numbers indicate thinner sheet steel. Thus the WR22 is the lightest deck—and probably the least expensive. You also should consider the width of the bottom of the rib. The wide bottom of the WR shape, for example, allows you to weld the deck to its supports. (This form of attachment is best for development of the deck diaphragm action for lateral forces.)

Regardless whether you choose the NR20, IR20, or WR22 units, the approximate deck weight is 2 psf, so the total dead load on the joists becomes 27 psf, not including the joist weight. Computations for the joists are listed in Figure 6.4. For the nearly flat roof surface, it is best to reduce joist deflection to a minimum; to do so, you use the deepest joists that are feasible. If the roof is placed on the top chord and the ceiling attached to the bottom chords, the deeper joists also provide more interstitial space for building services.

Since the ceiling should be flat, but the roof surface sloped, an accommodating joist is one with a flat bottom chord and a sloped top chord. For a conservative design in this event, use the depth at the shallow end to choose the actual joist size.

The light joist/trusses are very unstable in a direction perpendicular to the span, so you must provide considerable cross-bracing (called *bridging*), as recommended by the joist manufacturers. The ordinary means of support at the joist ends is by a seating of the extended end of the top chord (see Figure 6.1), which generally improves the lateral stability of the light trusses at the supports. In most situations, the bridging is required only to keep the bottom chords in place since the attachment of the deck keeps the top chords fixed in position against lateral effects.

A potential problem with the detail shown in Figure 6.1 is the bending induced in the wall by the eccentric load of the supported trusses. The only way to eliminate this bug is to support the trusses at the center of the wall, but

Computations for the Open Web Steel Joists

Joists at 6' centers, span of 60' (+ or -)

Loads:

DL = (27 psf X 6') = 162 lb/ft (not including weight of joist)

LL = (20psf X 6') = 120 lb/ft (not reduced)

Total load = 162 + 120 = 282 lb/ft + joist weight

Selection of Joist from Table B.16:

Note that the table value for total weight allowable includes the weight of the joist, which must be subtracted to obtain the load the joist can carry in addition to its weight.

Table data: total load = 282 lb/ft + joist, LL = 120 lb/ft, span = 60'

Options from Table B.16: None, deepest and heaviest table entry (30K12) carries only 262 lb/ft on 60' span.

Alternatives:

1. Decrease spacing of joists to lower unit load. At 5' centers the total load drops to 235 lb/ft. This is within the capacity of the 30K12, even if the joist weight of 17.6 lb/ft is added for an actual total load of 252.6 lb/ft. LL is also not a problem; Table allows 124 lb/ft at limit of L/360 deflection on the 60' span.

2. Table B.16 is abbreviated; other choices are available. Check selected manufacturer's tables for deeper, heavier joists.

3. The K series is the lightest series. Heavier, deeper joists are available in other series, with spans of 150' and more possible.

FIGURE 6.4 *Summary of computations for the steel roof structure.*

this method is not feasible with the parapet formed as shown in Figure 6.1. Figure 6.5 shows a construction that does place the truss supports at the center of the wall, but it involves considerable modification of the wall top and roof edge.

6.2 THE MASONRY WALLS

The form of construction for the masonry walls with CMUs is *structural reinforced masonry*. In structural reinforced masonry, you place steel reinforcing vertically and horizontally in the wall. The reinforcing goes in the voids in the hollow wall units, and these voids are filled with concrete. The resulting construction has a reinforced concrete frame inside its walls (see Figure 6.6).

Design codes require that the reinforced filled voids occur at least every 4 ft in both directions as well at wall tops and ends and at all edges of openings.

You can increase wall strength by various modifications, such as the following:

Use larger CMUs, producing a thicker wall.

Increase the size or number of reinforcing bars.

Increase the number of spaced reinforced vertical elements; you attain the greatest strength possible when all the CMU voids are filled and reinforced.

Provide individual strengthened portions (such as pilaster columns) within the wall construction for added stability, general increased strength, or response to concentrated loads.

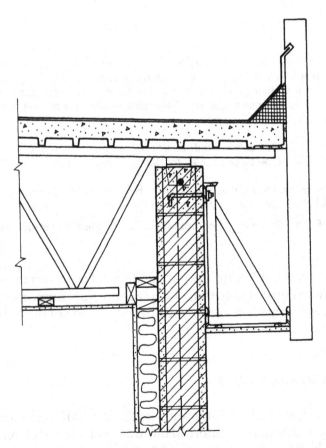

FIGURE 6.5 Variation of the roof-to-wall construction joint. For comparison, see Figure 6.1.

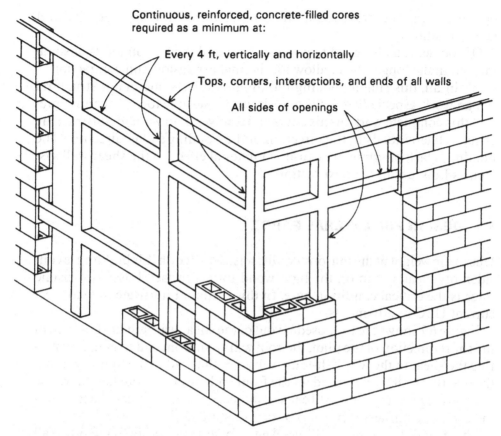

Continuous, reinforced, concrete-filled cores
required as a minimum at:

Every 4 ft, vertically and horizontally

Tops, corners, intersections, and ends of all walls

All sides of openings

FIGURE 6.6 Reinforced masonry, using hollow concrete blocks (CMUs).

In the roof structure shown in Figures 6.1 and 6.3, the end walls carry only a small roof load from the 6 ft span deck. Even if this load is applied eccentrically and thus induces some bending in the wall, the minimum wall construction should be sufficient at these locations. This load does not consider shear wall actions, of course.

The walls at the front and rear receive considerably greater gravity loads from the 60-ft-span trusses. If applied eccentrically to the wall as in Figure 6.1, these loads may well create a critical situation with combined compression and bending in the wall. If so, you need to increase wall strength or effectively eliminate the bending by modifying the construction to place the load from the trusses at the wall center (see Figure 6.5).

At 6 ft on center, the truss loads constitute a considerable concentration of force at the top of the wall. However, strengthening the horizontal beam at the top of the wall will distribute the load to the rest of the wall.

With all its openings, the front wall is really a set of individual elements, each requiring investigation. Header portions over openings must be designed as beams. Vertical edges at the sides of openings must be designed as columns for the support of the headers. Portions of wall remaining between openings must be designed as individual piers—for both gravity and lateral loadings.

Of course, you also need to provide structural separation joints at some interval in the long walls to allow for thermal expansion and contraction.

All in all, this simple-looking building requires some major structural design work, especially if it has to resist major seismic forces. I cannot present this complete work in this chapter. However, I do want to illustrate the effect of lateral loads in the direction of the building's long direction. The building's lack of symmetry causes a torsional effect on the shear wall system, as I discuss in the next section.

6.3 DESIGN FOR LATERAL FORCE

Due to the weight of the masonry walls, seismic force on this building is considerably greater than on the light wood frame. In fact, I assume seismic force is the critical condition here. In this section, I illustrate how to meet current UBC requirements.

Figure 6.7 shows the proposed layout of the shear wall system. For load in the short direction of the plan, the roof will span from end to end, transferring the shear to the two 45-ft-long walls. For the load in the long direction, the five 10 ft walls will be used on the front and the entire wall length will be used on the rear. The latter will cause some eccentricity between the load and the center of stiffness (centroid) of the resisting walls.

The dead loads to be used for seismic force are listed in Table 6.1. Wall loads are taken as the upper half of the walls, ignoring openings that are mostly in the lower half. The canopy load is taken by the roof in both directions. A nominal load is assumed for HVAC units on the roof. Although the code requires that a minimum torsion be considered by placing the load off center by 5% of the building's length, the shear walls will barely feel the effect.

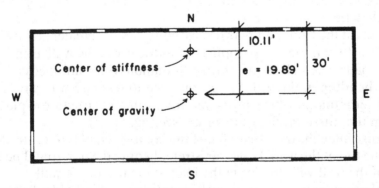

FIGURE 6.7 Layout of the shear walls. Center of gravity refers to the building weight. Center of stiffness indicates the centroid of the shear walls for torsional action. Eccentricity between the centers of gravity and stiffness indicates a potential for torsional effect due to lateral loads.

The lateral seismic shear force is determined as

$$V = \frac{ZIC}{R_W} W$$

where

Z = 0.4 (zone 4, *UBC* Table 16-I)
I = 1 (ordinary or standard occupancy, *UBC* Table 16-K)
C = 2.75 (maximum value, *UBC* Section 1628.2.1)
R_W = 6 (masonry shear wall, *UBC* Table 16-N
W = building weight for direction considered

TABLE 6.1   Loads to the Roof Diaphragm (kips)

Load Source and Calculation	North–South Load	East–West Load
Roof Dead Load		
145 × 60 × 20 psf	174	174
East and West Exterior Walls		
50 × 11 × 70 psf × 2	0	77
10 × 6 × 70 psf × 2	0	9
10 × 6 × 10 psf × 2	0	1
North Wall		
150 × 12 × 70 psf	126	0
South Wall		
65.3 × 10 × 70 psf	46	0
84 × 6 × 70 psf	35	0
84 × 6 × 10 psf	5	0
Interior North–South Partitions		
60 × 7 × 10 psf × 5	21	21
Toilet Walls		
Estimated 250 × 7 × 10 psf	17	17
Canopy		
South: 150 × 100 plf	15	15
East and west: 40 × 100 plf	4	4
Rooftop HVAC Units		
Estimated	5	5
Total load	448 [1993 kN]	323 [1434 kN]

Thus, for the two axes of the building,

$$V = \frac{(0.4)(1)(2.75)}{6} W = 0.1833 W$$

Using the data from Table 6.1

$$\text{N-S } V = (0.1833)(448) = 82 \text{ kips}$$

At the ends of the building, the maximum shear stress in the roof diaphragm due to the north–south seismic force is

$$\text{maximum } v = \frac{41,000}{60} = 683 \text{ plf}$$

This is a very high shear for the metal deck and would require a heavy-gauge deck and considerable welding at the diaphragm edge. Although it would be wise to reconsider the general design and possibly use at least one permanent interior partition, assume that the deck spans the building length for the shear wall design.

In the other direction, the shear in the roof deck is considerably less:

$$\text{E-W } V = 0.1833(323) = 59 \text{ kips}$$

$$\text{maximum } v = \frac{29,500}{150} = 197 \text{ plf}$$

This is very low shear for the deck, so if any interior shear walls are added, the deck gauge can probably be reduced to that required for the gravity loads only.

In the north–south direction, with no added shear walls, the end shear forces will be taken almost entirely by the long solid walls because of their relative stiffness. The shear force will be the sum of the end shear from the roof and the force due to the weight of the end wall.

$$
\begin{aligned}
\text{Wall weight} &= 18 \text{ ft} \times 50 \text{ ft} \times 70 \text{ psf} = 63,000 \text{ lb} \\
&\quad\ 7 \text{ ft} \times 10.67 \text{ ft} \times 70 \text{ psf} = 5228 \text{ lb} \\
&\ 11 \text{ ft} \times 10.67 \text{ ft} \times 70 \text{ psf} = \underline{1174 \text{ lb}} \\
&\qquad\qquad\qquad\qquad\quad \text{Total} = 69,402 \text{ lb}
\end{aligned}
$$

$$
\begin{aligned}
\text{Lateral force} &= 0.1833 \ W \\
&= 0.1833 \times 69.4 = 12.7 \text{ kips}
\end{aligned}
$$

The total force on the wall is thus $12.7 + 41 = 53.7$ kips, and the unit shear force in the wall is

$$v = \frac{53,700}{44.67} = 1202 \text{ lb/ft}$$

Assuming a 60% solid wall with 8 in. blocks, the unit stress on the net area of the wall is thus

$$v = \frac{1202}{12 \times 7.625 \times 0.60} = 21.9 \text{ psi}$$

If the reinforcing takes all shear and there is no special inspection, the allowable shear stress depends on the value of M/Vd for the wall:

$$\frac{M}{Vd} = \frac{(41 \times 15) + (12.7 \times 9)}{53.7 \times 44.67} = 0.304$$

Interpolating between the values for M/Vd of 0 and 1.0,

$$\text{allowable } v = 35 + (0.69)(25) = 52 \text{ psi}$$

You can increase this value by the usual one-third for seismic load to $(1.33)(52) = 69$ psi. So the masonry stress is adequate, but you must check the wall reinforcing for its capacity as shear reinforcement. With the minimum horizontal reinforcing determined previously—No. 5 at 48 in.—the load on the bars is

$$V = (1202) \left(\frac{48}{12} \right) = 4808 \text{ per bar}$$

and the required area for the bar is

$$A_s = \frac{V}{f_s} = \frac{4808}{26,667} = 0.18 \text{ in.}^2$$

Thus the minimum reinforcing is adequate. These walls will feel additional stress from torsion, so some increase in the horizontal reinforcing is advisable.

Overturn is not a problem for these walls because of their considerable dead weight and the natural tiedown provided by the doweling of the vertical wall reinforcing into the foundations. These dowels also provide the necessary resistance to horizontal sliding.

In the east–west direction, the shear walls are not symmetrical in plan, so you must determine the location of the center of rigidity and then the torsional moment. The total loading is reasonably centered in this direction, so assume the center of gravity to be in the center of the plan.

The following analysis is based on examples in the *Masonry Design Manual* (Ref. 7). The individual piers are assumed to be fixed at top and bottom; their stiffnesses are found from Table C.2. The stiffness of the piers and the total wall stiffnesses are determined in Figure 6.8.

For the location of the center of stiffness, use the values determined for the north and south walls:

$$\bar{y} = \frac{(R \text{ for S wall})(60 \text{ ft})}{\begin{array}{c}(\text{sum of the } R \text{ values}\\ \text{for the N and S walls})\end{array}} = \frac{(2.96)(60)}{17.57} = 10.11 \text{ ft}$$

The torsional resistance of the entire shear wall system is the sum of the products of the individual wall rigidities times the square of their distances from the center of stiffness (see Table 6.2). The torsional shear load for each wall is

$$V_W = \frac{Tc}{J} R_i$$

where

V_W = total shear force on an individual wall
T = torsional moment = $(V)(e)$
c = distance of the wall from the center of stiffness
R_i = resistance value for the wall
J = torsional moment of inertia, as determined by the sum of the Rd^2 values for all the walls

In the north–south direction, *UBC* Section 1628.5 requires that the load be applied with a minimum eccentricity of 5% of the building length—in this case, 7.5 ft. Although this adjustment produces less torsional moment than the east–west load, it is added to the direct north–south shear and therefore critical for the end walls. The torsional load for the end walls is

$$V_W = \frac{(82)(7.5)(75)(3.17)}{44,524} = 3.28 \text{ kips}$$

As mentioned previously, this torsional effect should be added to the direct shear of 49,500 lb for the design of these walls.

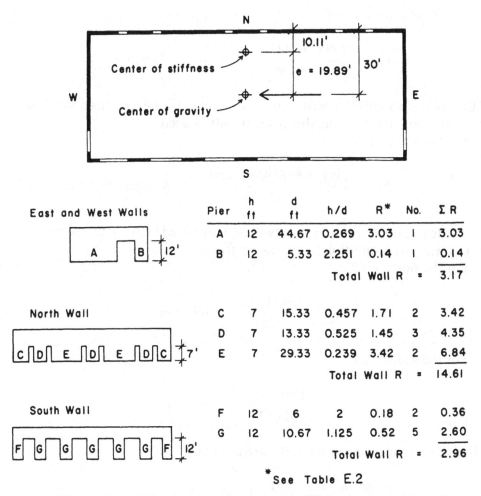

East and West Walls	Pier	h ft	d ft	h/d	R*	No.	Σ R
	A	12	44.67	0.269	3.03	1	3.03
	B	12	5.33	2.251	0.14	1	0.14
					Total Wall R	=	3.17

North Wall	Pier	h ft	d ft	h/d	R*	No.	Σ R
	C	7	15.33	0.457	1.71	2	3.42
	D	7	13.33	0.525	1.45	3	4.35
	E	7	29.33	0.239	3.42	2	6.84
					Total Wall R	=	14.61

South Wall	Pier	h ft	d ft	h/d	R*	No.	Σ R
	F	12	6	2	0.18	2	0.36
	G	12	10.67	1.125	0.52	5	2.60
					Total Wall R	=	2.96

*See Table E.2

FIGURE 6.8 Analysis of the masonry piers (shear walls) for torsional action due to lateral loads.

TABLE 6.2 Torsional Resistance of the Masonry Shear Walls

Wall	Total Wall R	Distance from Center of Stiffness (ft)	$R(d)^2$
South	2.96	49.89	7,367
North	14.61	10.11	1,495
East	3.17	75.00	17,831
West	3.17	75.00	17,831
Total torsional moment of inertia (J)			44,524

For the north wall:

$$V_W = \frac{(59)(19.89)(10.11)(14.61)}{44,524} = 3.89 \text{ kips}$$

(This load is actually opposed to the direct shear, but the code does not allow the reduction and thus the direct shear only is used.)

For the south wall:

$$V_W = \frac{(59)(19.89)(49.89)(2.96)}{44,524} = 3.89 \text{ kips}$$

The total direct east–west shear will be distributed between the north and south walls in proportion to the wall stiffnesses.

For the north wall:

$$V_W = \frac{(59)(14.61)}{17.57} = 49.1 \text{ kips}$$

For the south wall:

$$V_W = \frac{(59)(2.96)}{17.57} = 9.94 \text{ kips}$$

The total shear loads on the walls are therefore

north: $V = 49.1$ kips

south: $V = 3.89 + 9.94 = 13.83$ kips

The loads on the individual piers are then distributed in proportion to the pier stiffnesses (R) as determined in Figure 6.8. The calculation for this distribution and the determination of the unit shear stresses per foot of wall are shown in Table 6.3. A comparison with the previous calculations for the end walls will show that these stresses are not critical for the 8 in. block walls.

In most cases the stabilizing dead loads plus the doweling of the end reinforcing into the foundations is sufficient to resist overturn effects. The heavy loading on the header columns and the pilasters provides considerable resistance for most walls. The only wall not so loaded is wall C, for which the loading condition is shown in Figure 6.9. The overturn analysis for this wall is as follows:

$$\text{overturn } M = (5740)(7.0) = 40,180 \text{ ft-lb}$$

$$\text{stabilizing } M = (0.85)(23,000)\left(\frac{15.33}{2}\right) = 149,851 \text{ ft-lb}$$

TABLE 6.3 Shear Stresses in the Masonry Walls

Wall	Shear Force on Wall (kips)	Wall R	Pier	Pier R	Shear Force on Pier (kips)	Pier Length (ft)	Shear Stress in Pier (lb/ft)
North	49.1	14.61	C	1.71	5.74	15.33	374
			D	1.45	4.87	13.33	364
			E	3.42	11.50	29.33	392
South	13.83	2.96	F	0.18	0.84	6.00	140
			G	0.52	2.43	10.67	228

These computations indicate that the wall is stable without any requirement for anchorage even though the wall weight in the plane of the shear wall was not included in computing the overturning moment.

6.4 CONSTRUCTION DETAILS

In this section, I show some possible details for the structure for Building Two. The all-wood structure was shown in Chapter 5, while the structure with a steel roof was shown earlier in this chapter. Here I show a system with masonry walls and a wood roof. The roof system uses columns consisting of masonry pilasters at the exterior walls and steel pipe columns at the building interior; it also employs a beam-purlin-rafter-deck system.

Figure 6.10 shows partial plans for the roof structure and the foundation system. The basic components of the modular roof system are indicated. While you can use odd-sized or cut plywood panels, this system works just as well when you use the standard 4 ft × 8 ft panels commonly available from material suppliers.

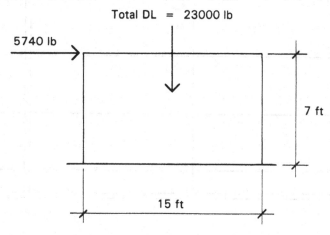

Total DL = 23000 lb

5740 lb

7 ft

15 ft

FIGURE 6.9 Loading for overturn investigation of the shear wall.

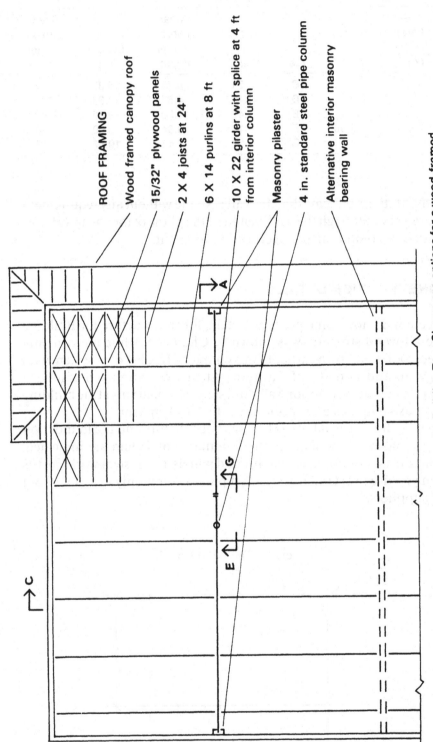

ROOF FRAMING

Wood framed canopy roof

15/32" plywood panels

2 X 4 joists at 24"

6 X 14 purlins at 8 ft

10 X 22 girder with splice at 4 ft from interior column

Masonry pilaster

4 in. standard steel pipe column

Alternative interior masonry bearing wall

FIGURE 6.10 Structural plans for Building Two. Shown are options for a wood-framed roof structure with interior columns and an interior bearing wall. Section marks refer to details in Figures 6.11 and 6.12.

82

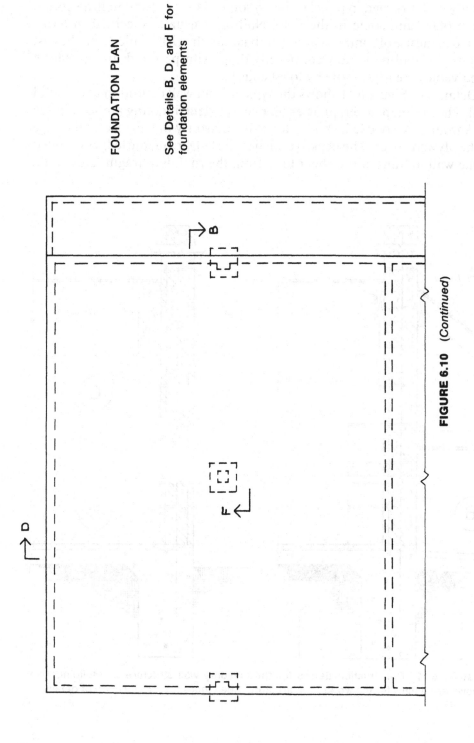

FOUNDATION PLAN

See Details B, D, and F for foundation elements

FIGURE 6.10 (*Continued*)

83

A nailing schedule or plan would show the variation of nailing required for the roof diaphragm actions. For this system the joists and plywood are usually prefabricated in panel units, so some of the plywood nailing occurs in the plant and some in the field. Nailing is usually specified in terms related to the use of common nails (the basis for the UBC tables), but the vast majority of nailing is now commonly done with powered drivers whose rated values are acceptable by local codes.

Detail A of Figure 6.11 shows the typical front wall condition at the solid wall. The girder, pilaster, pilaster pier, and widened footing are seen in the background. A wood ledger is bolted to the masonry wall to receive the edge of the plywood deck. The deck is nailed to the ledger and the ledger is bolted to the wall to transfer the shear load from the roof diaphragm to the wall.

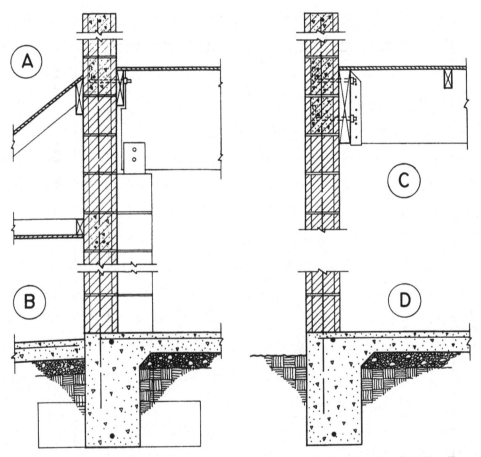

FIGURE 6.11 Construction details for the masonry wall structure for Building Two. Shown are details for the wood roof structure. See Figure 6.10 for location of details.

For the reinforced masonry wall, the code requires a minimum vertical spacing of solid horizontal reinforced bond courses. They also would be used at the top of the wall, the bottom of the header, and where the canopy and roof edge are bolted to the wall.

Detail B of Figure 6.11 shows the foundation edge at the front; note that it is essentially similar to that for the wood structure. The sill bolts would be replaced by dowels for the masonry wall. The pier would be added below the pilaster to carry the load down to the widened footing.

Detail C of Figure 6.11 shows the roof edge condition at the building ends. The wood ledger both provides support for the ends of the joists and transfers the lateral loads from the plywood deck to the masonry wall. Because of the roof slope, the top of the ledger varies 15 in. from the building's front to rear. The horizontal-filled block courses and the cutoff to a narrower parapet would be staggered to accommodate this slope. A somewhat-larger-than-usual cant could cover the jog in the wall to a narrower parapet block.

Detail D of Figure 6.11 is also reminiscent of the wood structure. If a footing of increased width is required, be careful to ensure that the centroid of the vertical loads is close to the center of the width of the footing.

Detail E of Figure 6.12 shows the typical form of the column-top support for a timber (or laminated) beam. Although this type of joint can resist some bending moment, the requirement in this situation is for vertical support only.

The detail could be developed with a wood column, for which the steel connector is probably available as a catalog item from some supplier. If a steel column is used, the U-shaped beam support would be welded to the top of the column.

This detail shows the support of a single beam that is continuous over the top of the column. A similar detail would be used if a beam joint occurs at the column top, although unbalanced loadings on the two beams would require more concern for the moment-resisting capability of the joint.

Detail F of Figure 6.12 shows the column base detail. For a wood column a common form of steel anchor/connector is one that effectively keeps the wood column from touching the concrete foundation. Again, this type of connector is a standard hardware item; it is available in a range of forms to accommodate various sizes of columns and magnitudes of vertical load support capacities.

The detail shows the use of a footing and short pier for support of the column, with the concrete floor slab separately developed. This configuration is particularly advantageous if the floor slab is to be cast after the roof structure is erected. Be sure to consider the desire for construction or control joints in the floor slab for crack control due to shrinkage or thermal change and the desire of providing or avoiding vertical support by the footing for the

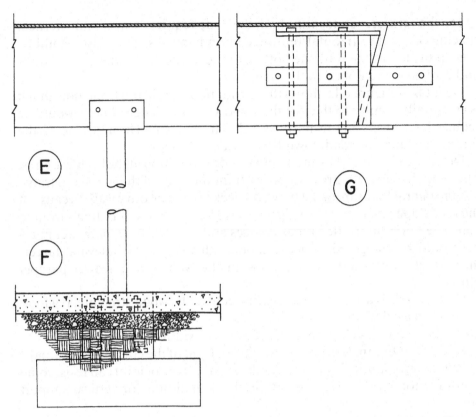

FIGURE 6.12 Construction details for the masonry wall structure for Building Two. Shown are details for the wood roof structure. See Figure 6.10 for location of details.

slab. Soil conditions, floor loadings, and various details of the interior construction can affect the development of these details.

Detail G of Figure 6.12 shows the form of a steel connector for one of the girder splices. In this case the end of the girder on the right is supported by the end of the girder on the left. You also may need to transfer tension and/or compression through the joint if the girder is required to perform collector or drag strut functions for lateral loads. This connector is also a standard hardware item.

7

DESIGN OF THE
STEEL TRUSS ROOF

If a gabled (double-sloped) roof form is desirable for Building Two, a possibile roof structure is shown in Figure 7.1. The building profile shown in Figure 7.1a shows a series of trusses, spaced at plan intervals, as shown for the beam-and-column rows in Figure 5.1c.

7.1 INVESTIGATION OF THE TRUSSES

The truss form is shown in Figure 7.1b, while the complete algebraic analysis for a unit gravity loading is displayed in Figure 7.2. And the true unit loading for the truss is derived from the form of construction described in Section 7.4 and tabulated in the next section.

The design illustrated here is based only on gravity loading. Although wind loading is important, the usual allowable stress increases for wind loads lessens its effects, except in regions of very strong windstorms. (The bending effects in the top chords are most likely to be increased if a deck is directly supported by the trusses.)

7.2 DESIGN OF THE STEEL TRUSS

Figure 7.1d shows the use of double-angle members with joints developed with gusset plates. The top chord is extended to form the cantilevered edge of the roof. For clarity sake, the detail shows only the major structural ele-

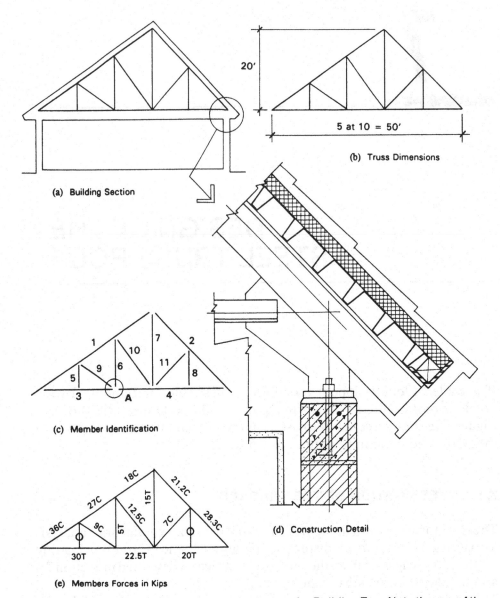

(a) Building Section

20'

5 at 10 = 50'

(b) Truss Dimensions

(c) Member Identification

(d) Construction Detail

(e) Members Forces in Kips

FIGURE 7.1   Form of the steel truss roof structure for Building Two. Note the use of the 50-ft span.

ments. Additional construction would be required to develop the roofing, ceiling, and soffit.

In trusses of this size, it is common to extend the chords without joints for as long as possible. Available lengths depend on the sizes of members and the lengths stocked locally. Figure 7.1c shows a possible layout; it creates a two-piece top chord and a two-piece bottom chord. The longer top chord piece is 36 ft plus the overhang, a length which may be difficult to obtain if the angles are small.

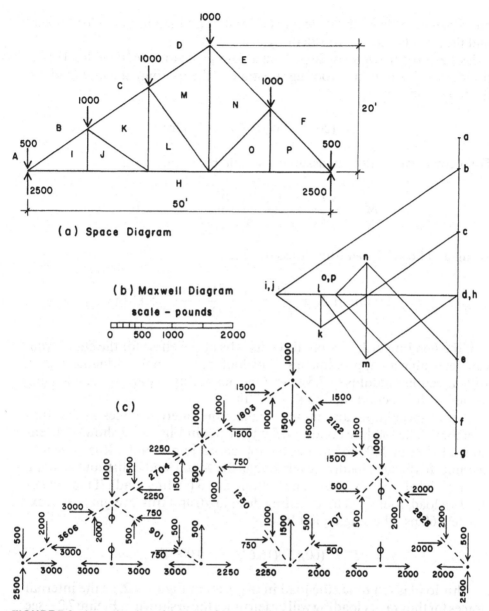

FIGURE 7.2 Investigation of the truss for gravity load. The space diagram in (a) indicates the truss layout and the unit loading. The Maxwell diagram (b) is a graphical analysis for the internal forces in the truss members. The separated joint diagram in (c) shows the individual joint force systems and the collective, resultant forces in the sloping truss members.

Figure 7.1d shows the use of a long-span steel deck bearing directly on top of the top chord. This option simplifies the framing by eliminating the need for intermediate framing between the trusses. For a truss spacing of 16-ft, 8-in., as shown in Figure 5.1c, the deck will be quite light and so this is a feasible system. However, the deck adds a spanning function to the top chord, and the chords will be considerably heavier.

Using a roof live load of 20 psf and assuming a total roof dead load of 25 psf (deck + insulation + roofing substrate + tile roofing), the unit load on the top chord is

$$w = (20 + 25)(16.67) = 750 \text{ lb/ft}$$

For a conservative design, assume a simple beam moment:

$$M = \frac{wL^2}{8} = \frac{(0.750)(10)^2}{8} = 9.375 \text{ k-ft}$$

Using an allowable bending stress of 22 ksi,

$$\text{Required } S = \frac{M}{F_b} = \frac{(9.375)(12)}{22} = 5.11 \text{ in.}^2$$

If the bending task is about three-fourths of the effort for the chord, you can determine an approximate size by looking for a pair of double angles with a section modulus of 7.5 or so. One possibility: a pair of $6 \times 4 \times \frac{1}{2}$-in. angles, with a section modulus of 8.67 in.^{2}.

Before speculating any more, calculate the internal forces in the truss members. The loading condition for the truss in Figure 7.2 shows concentrated forces of 1000 lb each at the top chord joints. Note: It is typical to assume this form of loading, even though the actual load is distributed along the top chord (roof load) and the bottom chord (ceiling load). If the total of the live load, roof dead load, ceiling dead load, and truss weight is approximately 60 psf, the single joint load is

$$P = (60)(10)(16.67) = 10,000 \text{ lb.}$$

This load is ten times the load in the truss in Figure 7.2, so the internal forces for the gravity loading will be ten times those shown in Figure 7.2—see Figure 7.1e.

Table 7.1 summarizes the design of all the truss members except the top chords. Tension members in the table were selected on the basis of welded joints and a desire for a minimum angle leg of $\frac{3}{8}$ in. thickness. Compression members were selected from AISC tables.

As in many light trusses, a minimum-sized member is established by the layout, dimensions, magnitude of forces, and the details of the joints. Angle legs wide enough to accommodate bolts or thick enough to accommodate fillet welds are such concerns. Minimum L/r ratios for members are another. Minimum requirements sometimes result in a poor truss design, when the combination of form, size, and proposed construction type do not mesh well.

TABLE 7.1 Design of the Truss Members

	Truss Member		
No.	Force (kips)	Length (ft)	Member Choice (All Double Angles)
1	36C	12.00	Combined bending and compression member $6 \times 4 \times \frac{1}{2}$
2	28.3C	14.2	Max. $L/r = 200$, min. $r = 0.85$ $6 \times 4 \times \frac{1}{2}$
3	30T	10.00	Max. $L/r = 240$, min. $r = 0.50$ $3 \times 2\frac{1}{2} \times \frac{3}{8}$
4	22.5T	10.00	$3 \times 2\frac{1}{2} \times \frac{3}{8}$
5	0	6.67	$2\frac{1}{2} \times 2\frac{1}{2} \times \frac{3}{8}$
6	5T	13.33	Max. $L/r = 240$, min. $r = 0.67$ $2\frac{1}{2} \times 2\frac{1}{2} \times \frac{3}{8}$
7	15T	20.00	Max. $L/r = 240$, min. $r = 1.00$ $3\frac{1}{2} \times 3\frac{1}{2} \times \frac{3}{8}$
8	0	10.00	$2\frac{1}{2} \times 2\frac{1}{2} \times \frac{3}{8}$
9	9C	12.00	$2\frac{1}{2} \times 2\frac{1}{2} \times \frac{3}{8}$
10	12.5C	16.67	Max. $L/r = 200$, min. $r = 1.00$ $3\frac{1}{2} \times 2\frac{1}{2} \times \frac{3}{8}$
11	7C	14.20	Min. $r = 0.85$ $3 \times 2\frac{1}{2} \times \frac{3}{8}$

The truss chords could consist of structural tees, with most gusset plates eliminated, except possibly at the supports. If the trusses can be transported to the site and erected in one piece, all the joints within the trusses could be shop-welded. Otherwise, you must develop a scheme for dividing the truss and splicing the separate pieces in the field; most likely, you need to use high-strength bolts for the field connections.

To derive an approximate design of the top chord, consider a combined function equation of the simple form of the straight-line interaction graph:

$$\frac{f_a}{F_a} + \frac{f_b}{F_b} = 1$$

To implement this equation, consider two ratios: the compression required to the compression capacity and the actual bending requirement (required S) to the actual S. Thus

Required compression = 36.06 kips
Capacity (Load table in Ref. 2) = 137 kips

Ratio $= 36.06/137 = 0.263$

Required $S = 5.11$ in.^{3}

Actual S (Property Table in Ref. 2) $= 8.67$ in.^{3}

Ratio $= 5.11/8.67 = 0.589$

And, the sum of the ratios $= 0.852$

These computations indicate that the proposed choice is reasonable, although a more elaborate investigation is required by the AISC specifications.

A possible configuration for Joint A (see Figure 7.1b) is shown in Figure 7.3a. Since the bottom chord splice occurs at this point (see Figure 7.1c), the chord is shown as discontinuous at the joint. If the truss is site-assembled in two parts, this joint might accommodate field connecting; the alternative is shown in Figure 7.3b, with the splice joint achieved with bolts.

7.3 DEVELOPMENT OF THE ROOF STRUCTURE

Development of the roof structure involves the design of a roof deck system and probably a framed ceiling system. These may be directly supported by the truss chords or may begin with an intermediate set of framing members supported at the truss joints. The general framing plan for both options is shown in Figure 7.4.

When trusses are closely spaced (as with the open web joists in Chapter 6), the system shown in Figure 7.4a is most probable. For the scheme in this chapter, the truss spacing is marginally too wide to utilize a clear-spanning deck—although such decks are available. For the other option, the projected

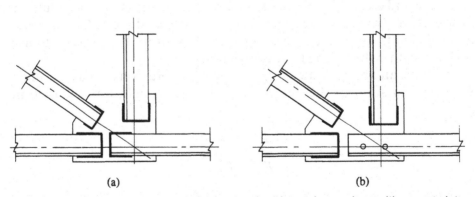

(a)   (b)

FIGURE 7.3 Form of a typical truss joint using double-angle members with gusset plates and welded connections. (a) Indicates a fully welded joint. (b) Indicates a possibility for a field splice joint using steel bolts.

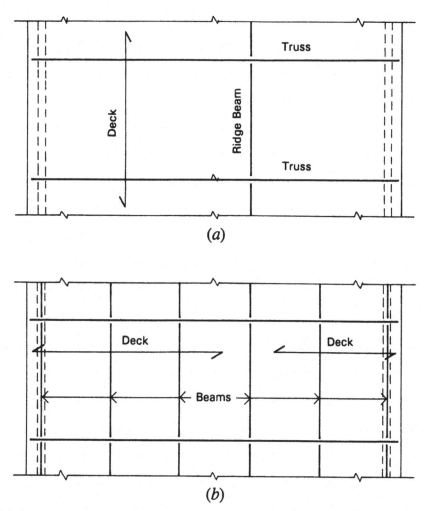

FIGURE 7.4 Partial framing plans for the truss roof structure. (a) With a structural deck spanning between trusses and no secondary framing. (b) With beams supported at truss joints and a deck supported by the beams.

horizontal span of the deck is only 10 ft, and a somewhat lighter deck is possible, but a whole new set of elements is required to span between the trusses and carry this deck. Compensating for the additional framing, however, is some elimination of bending in the truss chords.

Fire codes permitting, these trusses may be exposed to view on the interior, with no suspended ceiling below the trusses. However, it is likely that some structural supports would be required for lighting, fire sprinkler piping, HVAC ducts and fixtures, or other elements. You can support these elements in many ways, including direct support by truss members or wire hangers inserted through the roof deck.

PART III

BUILDING THREE

8

GENERAL CONSIDERATIONS
FOR BUILDING THREE

Building Three is a low-rise structure developed for commercial occupancy. This building is similar to Building Two—especially as an all-wood structure—and so it is shown with a similar plan. This similarity is pursued most directly in Chapter 9. Variations are developed in Chapters 10 and 11.

8.1 THE BUILDING

As shown in Figure 8.1, Building Three consists essentially of stacked Building Two producing a two-story building. The profile section of the building shows that the second-floor structure is developed essentially the same as the roof structure for Building Two in Figure 5.1*b*. For the roof, however, a clear span structure could be provided by 50 ft span trusses, as described in Section 5.5.

The jump from one to more stories involves some issues not pertinent to the single-story building:

Need for a framed floor. Although a single-story building with a basement or crawl space may need a framed floor, it is not inherent—as demonstrated with Building Two, which uses a concrete slab on grade for its floor.

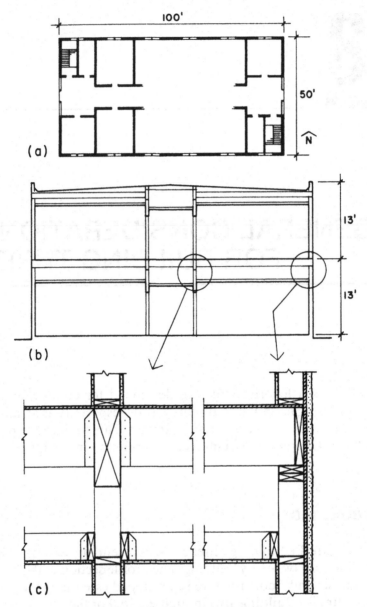

FIGURE 8.1   Building Three: general form.

Accommodation of vertical movement of people, materials, water, electric power, and so on. A multi-story building requires construction for stairs, elevators, pipe and wiring chases, duct shafts, and so on.

Need for stacking of vertical supports (columns and/or bearing walls); preferably on top of each other.

Increasing magnitudes of foundation loads, more or less in proportion to the number of stories.

More complex fire code requirements.

Although the second-floor structure for Building Three is similar to the roof structure for Building Two, they must accommodate different magnitudes of live load. The usual building code requirement for minimum load for offices is 50 psf (see Table A.3). However you also usually must add an allowance of 20 or 25 psf for movable partitioning. With a total live load of 75 psf, the design load will be several times greater than that for the roof. The difference in results is demonstrated in the next section.

The two-story building also will sustain a greater total wind force. The second-story shear walls will essentially be the same as those for Building Two, but the first-story shear walls will need to be stronger.

Another difference is that the floor structure will be dead flat, thereby eliminating the concerns that accompany a sloped surface. Note that the details for the second-floor framing, shown Figure 8.1c, indicate the use of platform framing. Otherwise the details of the construction are similar to those for Building Two except for the possible use of the trusses for the roof. One way to economize with the trusses is to eliminate a general framing for the ceiling and attach the ceiling surface directly to the bottom chords of the trusses. And you can slope the top chords of the trusses to provide the framing for the roof deck in the profile necessary for drainage. The planning of the building, the location of roof drains, and the framing plan for the trusses must be co-ordinated to bring these simplifications off effectively.

8.2 STRUCTURAL ALTERNATIVES

For buildings of limited height, the all-wood structure remains a reasonable alternative, depending on the total floor area and any special restrictions for fire zones. However, many other forms of construction are also possible—running the gamut of steel, concrete, masonry, and various mixtures. Economics, local traditions, building owner's preferences, and architectural design decisions will vary often enough to guarantee a wide variety of examples.

I describe three solutions in this part; however, the two solutions developed for Building Four are also possibilities, along with many others. As building height and total floor area increase, the only two alternatives tend to be those illustrated for Building Four.

In any region a few forms of construction are popular, at any give time. When economics, time for construction, and general ease of design are critical concerns, these popular forms become the standards by which more imaginative solutions will be evaluated. If you want to push aside the common solutions, be prepared for some finely tuned analysis of your designs.

The solutions illustrated here in Chapter 9 and in Chapters 12 and 13 of Part Four are pretty common.

9

DESIGN OF THE
LIGHT WOOD STRUCTURE

This chapter presents a development of Building Three, using the all-wood structure illustrated in Figure 8.1.

9.1 THE FRAMED FLOOR STRUCTURE

The increased loads require that the member sizes for the framing of the floor be considerably greater than those needed for the roof for Building Two. (Here I illustrate the design with structural lumber; I discuss alternatives later.)

You can assume the following for the dead load of the floor construction, as long as the ceiling is separately supported (that is, not hung from the floor joists):

Carpet and pad	3.0 psf
Fiberboard underlay	3.0
Plywood deck, $\frac{1}{2}$ in.	1.5
Ducts, lights, wiring	3.5
Total without joists:	11.0 psf

With joists at 16 in. spacing, the loading for a single joist is

$$DL = \frac{16}{12} \ (11) \ = \ 14.6 \ \text{lb/ft} \ + \ \text{the joist, say 20 lb/ft}$$

$$LL = \frac{16}{12}(75) = 100 \text{ lb/ft}$$

$$\text{Total} = 120 \text{ lb/ft}$$

For the 21 ft span joists, the maximum bending moment is

$$M = \frac{wL^2}{8} = \frac{(120)(21)^2}{8} = 6615 \text{ ft-lb}$$

For select structural Douglas Fir–Larch joists with 2 in. nominal thickness, F_b from Table B.4 is 1668 psi for repetitive member use. Thus the required section modulus is

$$S = \frac{M}{F_b} = \frac{(6615)(12)}{1668} = 47.6 \text{ in.}^3$$

Table B.3 shows that this section modulus is just over the value for a 2 × 14, which is the largest member listed with a 2 in. nominal thickness. You can increase the stress grade of the wood, use a thicker joist, or reduce the joist spacing. If you reduce the spacing to 12 in., the required section modulus drops by one-fourth, making the 2 × 14 adequate for flexure.

Shear is unlikely to be critical with long-span joists, but you should investigate deflection. The usual deflection limit for this situation is a maximum live load deflection of $\frac{1}{360}$ of the span, or (21)(12)/360 = 0.7 in. With the 2 × 14 at 12 in. spacing, the maximum deflection under live load only is

$$\Delta = \frac{5}{384}\frac{wL^4}{EI} \text{ or } \frac{5}{384}\frac{WL^3}{EI}$$

$$= \frac{5}{384}\frac{(75 \times 21)(21 \times 12)^3}{(1,900,000)(291)} = 0.60 \text{ in.}$$

so deflection is not critical.

The beams support not only the 21 ft joists, but also the 8 ft joists at the corridor. The corridor live load is usually required to be 100 psf, but the short span probably permits a small joist—usually a 2 × 6 minimum—so the corridor dead load is slightly less. Also the total area supported by a beam exceeds 150 ft², which allows for a minor reduction of the live load (see Section A.2). To simplify the computations, use the dead and live loads as determined for the 21 ft span joists as the design load for the beam for the total of 14.5 ft of joist span (half the distance to the other beam plus half the distance to the exterior wall). Thus the beam load is

$$DL = (16)(14.5) = 232 \text{ lb/ft} \quad \text{(joist load)}$$

$$+ \text{ beam weight} = 30 \quad (\text{estimate})$$
$$+ \text{ wall above} = \underline{150} \quad (\text{2nd-floor corridor})$$
$$\text{Total } DL = 412 \text{ lb/ft}$$
$$LL = (75)(14.5) = 1088 \text{ lb/ft}$$
$$\text{Total load} = 412 + 1088 = 1500 \text{ lb/ft}$$

For a uniformly loaded, simple span beam with a span of 16 ft 8 in., determine

$$\text{Total load} = W = (1.5)(16.67) = 25 \text{ kips}$$
$$\text{End reaction and maximum shear} = W/2 = 12.5 \text{ kips}$$
$$\text{Maximum moment} = wL^2/8 = (1.5)(16.67)^2/8 = 52.1 \text{ kip-ft}$$

For a Douglas Fir–Larch, dense No. 1 grade beam, Table B.4 yields values of $F_b = 1550$ psi, $F_v = 85$ psi, and $E = 1,700,000$ psi. To satisfy the flexural requirement, the required section modulus is

$$S = \frac{M}{F_b} = \frac{(52.1)(12)}{1.550} = 403 \text{ in.}^3$$

Table B.3 shows that the lightest section that will satisfy this requirement is an 8×20 with $S = 475$ in.^{3}

If the 20-in.-deep section is used, its effective bending resistance must be reduced. Thus the actual moment capacity of this section is reduced by using the size factor and is determined as

$$M = C_F \times F_b \times S = (0.947)(1.550)(475)(1/12) = 58.1 \text{ k-ft}$$

As this moment capacity exceeds the requirement, the correction for size effect is not critical when choosing the section.

If the actual beam depth is 19.5 in., the critical shear force may be reduced to that at a distance of the beam depth from the support. Thus you may subtract an amount of the load equal to the beam depth times the unit load. The critical shear force thus is

$$V = (\text{actual end shear}) - (\text{unit load times beam depth})$$

$$= 12.5 \text{ kips} - (1.5)\left(\frac{19.5}{12}\right) = 12.5 - 2.44 = 10.06 \text{ kips}$$

If you choose the 8×20, the maximum shear stress is

$$f_v = \frac{3}{2}\frac{V}{A} = \frac{(3)(10,060)}{(2)(146.25)} = 103 \text{ psi}$$

Even with the reduction in the critical shear force, this shear stress clearly exceeds the allowable stress of 85 psi. Thus you must increase the beam section to a 10×20 to satisfy the shear requirement. With this section, the flexural stress is reduced; so you may not need to use the dense grade of wood because the shear stress is a constant for all grades of the wood.

For beams of relatively short span and heavy loading, shear is a common controlling factor. As a result, you usually can't use a solid-sawn timber section, for which allowable shear stress is quite low. It is logical to modify the structure to reduce the beam span or to simply choose a steel beam or a glue-laminated section in place of the solid timber.

Although deflection is often critical for long-span, lightly loaded joists, it is seldom critical for the short-span, heavily loaded beam. You can verify this by investigating the deflection of this beam.

9.2 THE WALLS AND COLUMNS

Vertical supports for this structure are combinations of stud-framed bearing walls and solid-sawn wood columns. As I discussed for the exterior walls for Buildings One and Two, you must consider many variables to determine the exact details of the wall construction. For example, you may choose stud widths that define the interior, hollow spaces in the walls for other than structural reasons. Nonetheless your choice must withstand both vertical compression due to gravity and lateral bending due to wind or seismic forces.

The exterior walls are also used as shear walls for this building (see Section 9.3). This function may affect your choice of materials, the size of members, and various construction details, as well as your general planning of the walls. You must not only adequately distribute shear walls in the general building plan, but also be sure that individual walls have the proper form and dimensions.

Although some of the interior walls can be developed as bearing walls, this solution uses individual columns positioned at the corridor walls, permitting easy redevelopment of the building plan.

For the interior column at the first story, the design load is approximately the same as the total load on a single beam—that is, 25 kips. For the 10-ft-high column, Figure B.1 indicates an 8×8 column for a solid-sawn section. But it may be more practical to use a steel round pipe or a square tubular section whose size can fit into a 2×4 stud wall.

Columns must be provided at the beams' ends in the east and west walls. In the example these locations are the ends of the shear walls, and the normal use of a doubled stud at this point should result in an adequate column. If a beam end occurs in the center portion of a wall, you should provide a separate column, or special doubled stud.

9.3 DESIGN FOR WIND

The general design for wind includes the considerations enumerated at the beginning of Section 5.4. Investigation of the second-story studs in the exterior walls is similar to that made for Building One. At the first story in Building Three, the studs carry considerably more axial compression, but the bending due to wind is approximately the same as at the second story. The 2×6 studs at 16 in. centers are probably adequate at the first story. If an investigation shows an overstress condition, you can reduce the stud spacing to 12 in. or use a higher grade wood.

For lateral load the roof deck in Building Three is basically the same as that in Building Two, with the trusses it may be more practical to use an unblocked deck. Study the footnotes to Table B.7 with regard to the pattern of the layout of the plywood panels, especially for unblocked decks. Various special deck panels with tongue-and-groove edges are available in thicknesses greater than $\frac{1}{2}$ in., thus permitting truss spacings up to 4 ft.

The wind loading condition for the two-story building, shown in Figure 9.1a indicates a loading to the second-floor diaphragm of 235 lb/ft. With a

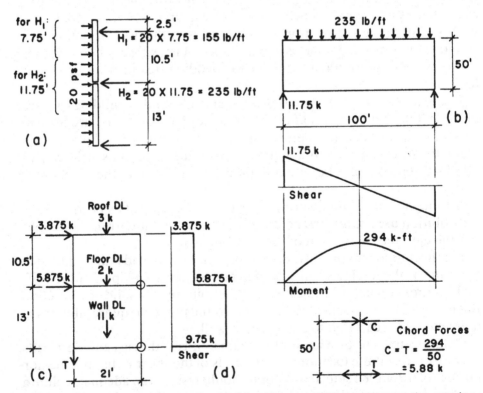

FIGURE 9.1 Investigation for wind on Building Three. (*a*) Wind loading on the exterior wall. (*b*) Actions of the second-floor horizontal diaphragm. (*c*) Actions of the two-story shear wall.

$\frac{15}{32}$-in.-thick deck as a minimum, the shear in the 50-ft-wide deck will not be critical. However, the stair wells at the east and west ends reduce the actual diaphragm width at the ends to only 35 ft. Figure 9.1*b* shows the loading for the second-floor deck and the critical shear and moment values for the diaphragm actions. At the ends the critical unit shear in the deck is

$$v = \frac{11,750}{35} = 336 \text{ lb/ft}$$

From Table B.7, you determine that this shear requires more than the minimum nailing. Options include:

1. Using $\frac{15}{32}$ in. Structural II deck with 8d nails at 4 in. at the diaphragm boundary and other critical edges.
2. Using $\frac{15}{32}$ in. Structural I deck with 10d nails at 6 in. throughout and 3× framing.
3. Using $\frac{19}{32}$ in. Structural II deck with 10d nails at 6 in. throughout and 3× framing.

At 8 ft from the building ends, the deck resumes its full 50 ft width, and the unit shear drops to

$$v = \frac{9870}{50} = 197 \text{ lb/ft}$$

Since this value is well below the capacity of the $\frac{19}{32}$ in. Structural II deck with minimum nailing, you may want to elect option 1, which involves 4 in. nailing in approximately 12% of the second-floor deck.

The diaphragm chord force for the second-floor deck is approximately 6 kips and must be developed in the framing at the wall, as shown in Figure 8.1*c*. You probably will want to use the continuous edge member at the face of the joists. The only real design problem is figuring out how to splice the member that will be made up of several pieces in the 100 ft length. Splicing may be achieved in a number of ways, but a joint using a steel strap with wood screws or a joint with bolts and steel plates should intrude the least in the construction.

The loading for the two-story end shear wall is shown in Figure 9.1*c*, and the shear diagram for this load is shown in Figure 9.1*d*. The second-story wall is similar to the end wall in Building Two (Section 5.4). Because minimal construction is adequate here, and no anchorage for overturn is required, the only problem is developing sliding resistance to the lateral force of 3875 lb. Since this wall does not sit on a concrete foundation, you must consider means of anchorage other than steel anchor bolts.

The lateral force in the second-story wall must be transferred to the first-story wall. Essentially this occurs directly if the plywood is continuous past the construction at the level of the second floor, as shown in Figure 8.1c. A critical location for stress transfer is at the top of the second-floor joists. At this point the lateral force from the second-floor deck is transferred to the wall through the continuous edge member. Therefore the nailing and plywood requirements for the first-story wall begin at this location. The last point for the nailing and plywood requirements for the second-story wall is at the sill for the second-story wall (on top of the second-floor deck, as shown in Figure 8.1c). The plywood for the wall and its nailing from this point down must satisfy the requirements for the first-story wall.

In the first-story wall, the total shear force is 9750 lb and the unit shear is

$$v = \frac{9750}{21} = 464 \text{ lb/ft}$$

If the $\frac{3}{8}$ in. Structural II plywood selected for Building Two is used for the second floor (see Section 5.4), it may be practical to use the same plywood for the entire two-story wall and to simply increase the nail size and/or reduce the nail spacing at the first story. Table B.8 yields a value of 410 lb/ft for $\frac{3}{8}$ in. Structural II plywood with 8d nails at 3 in. If the conditions of table footnote 3 are met, this value can be increased by 20% to 492 lb/ft. Although you have other options, this choice meets the lateral design criteria.

At the first-floor level, the investigation for overturn of the end shear wall is as follows (see Figure 9.1c):

$$\text{overturning moment} = (3.875)(23.5)(1.5) = 136.6 \text{ k-ft}$$
$$+ (5.875)(13)(1.5) = \underline{114.6 \text{ k-ft}}$$
$$\text{Total} = 251.2 \text{ k-ft}$$
$$\text{restoring moment} = (3 + 2 + 11)(21/2) = 168 \text{ k-ft}$$
$$\text{Net overturning moment} = 83.2 \text{ k-ft}$$

which requires an anchorage force at the wall ends of

$$T = \frac{83.2}{21} = 3.96 \text{ kips}$$

Since the safety factor of 1.5 for the overturn has already been used in the computation, it is reasonable to consider reducing this anchorage requirement to 3.96/1.5 = 2.64 kips if used in the form of a service load. In addition, the wind loading permits an increase of one-third in allowable stress, which you may also use to reduce the anchorage requirement. Finally, there are added dead load resistances at both ends of the wall: at the corridor the beam sits on the end of the wall, and at the corner this wall is firmly attached to the wall around the corner. Thus you may not need an anchorage device at all. However, most structural designers add a device as precaution.

10

DESIGN OF THE TIMBER AND MASONRY STRUCTURE

This chapter presents a solution for a type of Building Three that utilizes a form of construction known as *mill construction.* In this chapter, I present some typical mill construction details as they appeared in popular early twentieth century books on building technology.

Mill construction evolved during the eighteenth and nineteenth centuries as more industrial and commercial buildings were built. A common mill construction form in the late nineteenth and early twentieth centuries consisted of exterior walls of structural brick masonry, interior floor construction of heavy timber, and roof structures using trusses of steel or combinations of timber and steel (see Figures 10.1 through 10.5).

Figure 10.6 shows the general form of Building Three, as adapted for mill construction. Note that a third story has been added; a truss provides a column-free space on the top floor. Some of the details for this construction are shown in Figure 10.6. Additional details are presented later in this chapter.

10.1 TIMBER FLOORS AND COLUMNS

The system for the interior construction, shown in the upper illustration in Figure 10.3, uses heavy timber sections for the columns and main beams and a thick timber deck. Maintaining minimum thicknesses of the principal structural elements provides for a limited hourly fire rating for the construc-

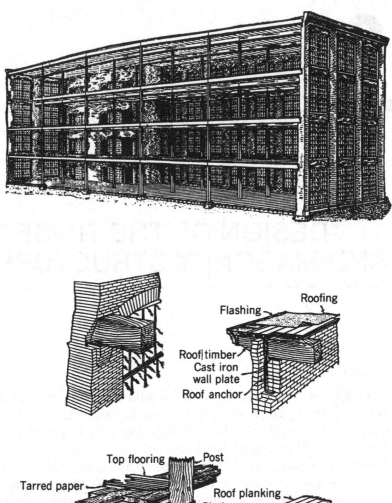

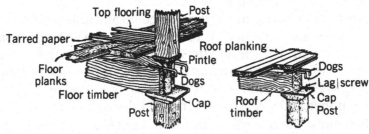

FIGURE 10.1 Example of mill construction for a multistory building. Reproduced from *Construction Revisited* (Ref. 20) with permission of the publisher, John Wiley & Sons.

tion (classified as Type IV—Heavy Timber in the *Uniform Building Code*). If you assume that the system can be exposed on the building interior, the only covering in general is that for the floor surfaces: you can use various finish floor materials (hardwood strip, vinyl tile, carpet, and so on) over the structural plywood that is nailed on top of the timber deck (see Figure 10.6).

Figure 10.7 shows a general framing plan for the typical upper floor system. This system is similar to the light wood frame solution in Chapter 9, except for the spacing of the beams that carry the deck.

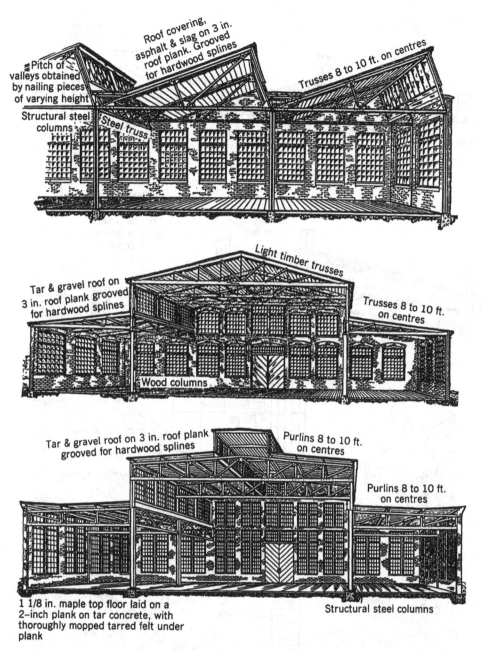

FIGURE 10.2 Examples of mill construction for one-story industrial buildings with different roof forms. Reproduced from *Construction Revisited* (Ref. 20) with permission of the publisher, John Wiley & Sons.

In the past, this system's columns, beams, and thick deck would all consist of solid-sawn lumber elements. The illustration in Figure 10.3 shows a possible modification, in which bundled thinner elements are used to obtain the larger beams.

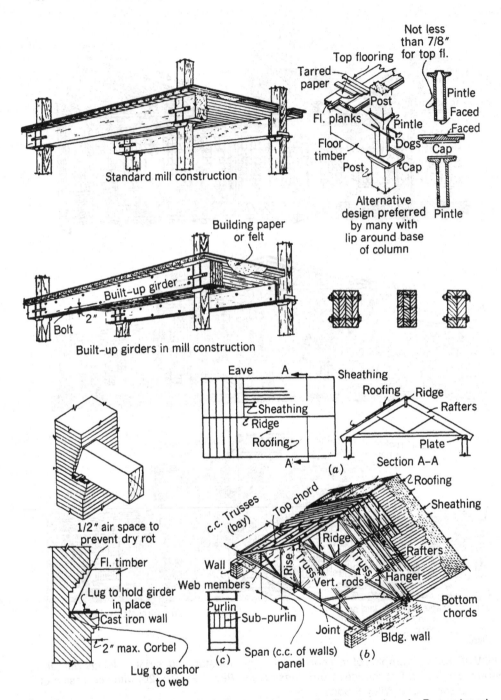

FIGURE 10.3 Examples of heavy timber construction for floors and roofs. Reproduced from *Construction Revisited* (Ref. 20) with permission of the publisher, John Wiley & Sons.

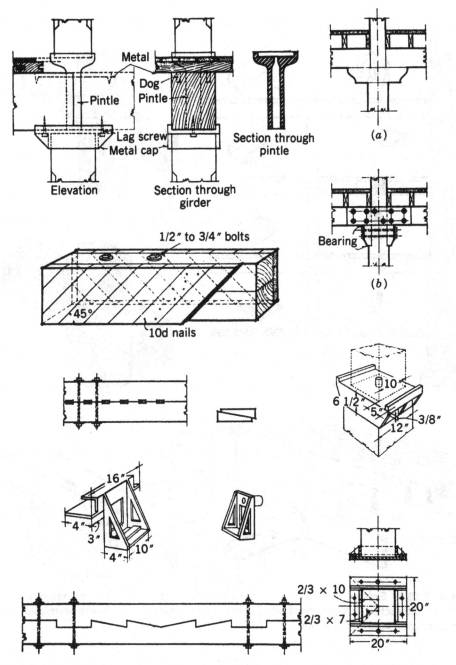

FIGURE 10.4 Typical elements of early twentieth century heavy timber construction. Reproduced from *Construction Revisited* (Ref. 20) with permission of the publisher, John Wiley & Sons.

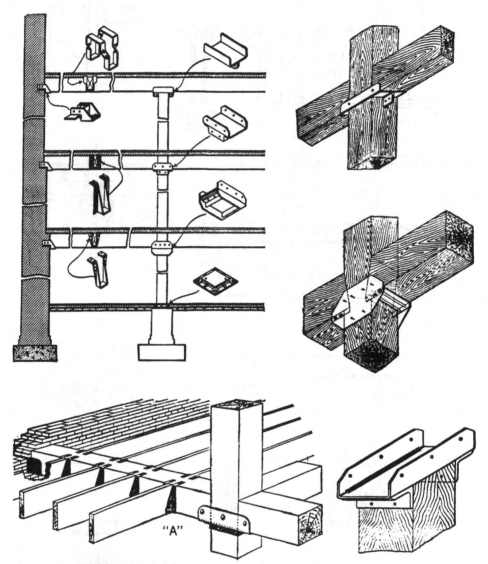

FIGURE 10.5 Examples of early twentieth century steel framing devices for masonry and timber construction. Reproduced from *Construction Revisited* (Ref. 20) with permission of the publisher, John Wiley & Sons.

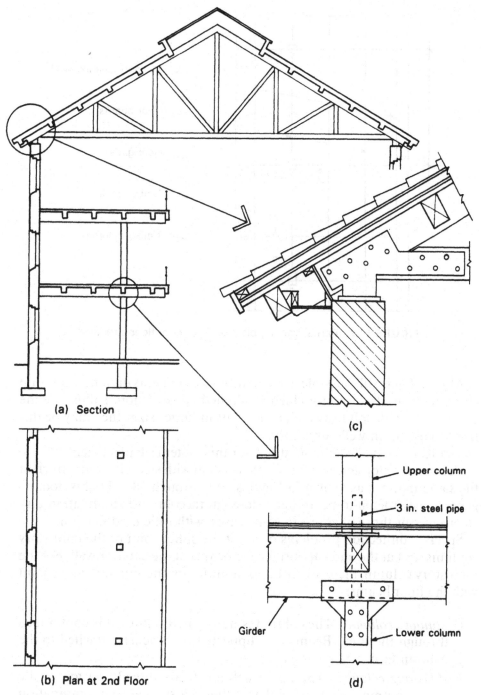

(a) Section

(c)

Upper column

3 in. steel pipe

Girder

Lower column

(b) Plan at 2nd Floor

(d)

FIGURE 10.6 General form of the timber and masonry structure for Building Three.

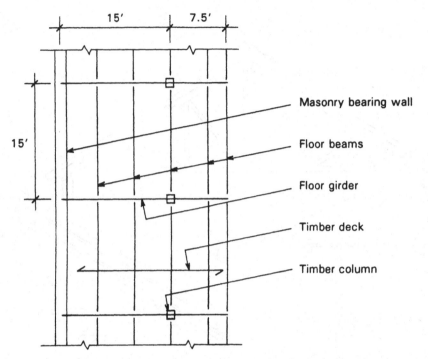

FIGURE 10.7 Partial framing plan for the timber floor structure.

Although it is still possible to use solid-sawn elements or bolted, built-up elements, you also can use glued laminated construction. I illustrate the design here with solid-sawn elements, but in larger sizes they may be difficult to obtain in well-cured form.

Figure 10.8 lists the computations for this system's typical elements.

You can develop connections for this system with steel elements, much in the same manner as shown in Figures 10.1 through 10.5. That is, today's general tasks (beam-to-beam, beam-to-column, column-to foundation, and so on) mirror those of yore—in some cases, with little modification.

Special connection problems occur at the splice joint of the multistory columns and at the end support for the beams at the exterior wall. For the multistory column, there are three possibilities for the situation at the joint with the floor beams:

Continuous columns. The column is not jointed; rather, it is continuous through the joint. Beams are supported on brackets attached to the column face.

End-bearing columns. The upper column bears directly on top of the lower column, using a steel device that is a single-piece combination base and cap in one. Beams are supported by the same steel cap/base or on wood bolsters, as shown in Figure 10.4.

Pintles. This device generally permits the beam to pass through the joint while keeping the column from actually bearing on top of the beam. See Figures 10.4 and 10.5.

FLOOR DECK

Assume timber deck + plywood + carpet & pad = 15 psf DL.

Use LL = 100 psf (office + partitions, or corridor).

Select commercial deck, probably 2" nominal thickness.

BEAM, 5 ft c/c, 15 ft span

Total load = 5 X 15 X 115 = 8625 lb, add total of 300 lb for beam, use 8925 lb.

M = WL/8 = (8925)(15)/8 = 16,734 ft-lb or 200,813 in.-lb.

Douglas fir-larch, No. 1, from Table B.4: allowable bending stress = 1300 psi.

Required S = M/F_b = 200,813/1300 = 155 in.³, Table B.3: use 8 X 12, S = 165 in.³.

Shear: V = 8925/2 = 4463 lb, f_v = 1.5V/A = 1.5(4463)/86.25 = 77.6 psi, not critical.

Deflection, see Fig. B.2, 12 in. depth not critical for LL defl. of L/360.

GIRDER, loads and span as shown

Area supported = 15 X 22.5 = 337.5 ft², use red. LL, approx. unit beam load = 8 k.

Max. M = 70 k-ft, or 840 k-in.

Required S = 840/1.3 = 646 in.³.

Use 10 X 22 or 12 X 20.

Max. V = 12.7 kips,
f_v = 1.5(12,700)/204 = 93 psi for 10 X 22.

f_v = 1.5(12,700)/224 = 85 psi for 12 X 20 (just adequate).

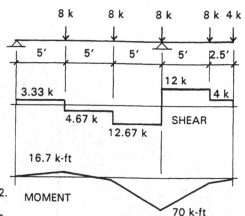

COLUMN, 10 ft high to bottom of girder

In 1st story, total load = approx. 2 X (girder reaction) = 66 kips.

Figure B.1: 10 X 10 required (or 8 X 12).

FIGURE 10.8 Summary of computations for the timber structure.

Pintles are not used much any more, but either of the other options is possible. However, be sure to consider how long a timber section you can obtain: the shorter the required length, the easier it is to find a good piece of timber.

10.2 TIMBER ROOF TRUSS

As shown in Figure 10.6, the roof structure consists of a clear-spanning truss. There are several options for the layout, materials, and assemblage details of this truss. In Figure 10.2 the upper and lower figures show the use of a light steel truss with joints achieved with gusset plates—a common form then and now for moderate spans.

The middle figure in Figure 10.2 shows the use of a combination timber and steel truss. This form was common because it eliminated most tension connections between wood members. Figure 10.9 shows more detailed illustration of this truss form, which is the basic form used in this chapter.

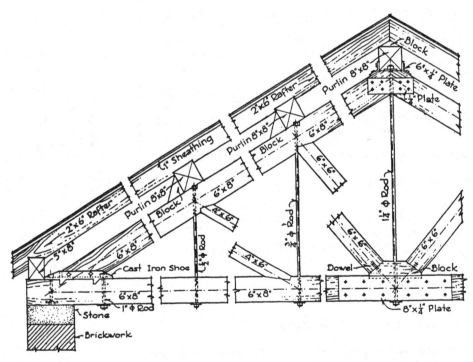

FIGURE 10.9 Details of an early twentieth century timber truss. (From *Materials and Methods of Construction* by Gay and Parker, 1932, reproduced with permission of the publisher, John Wiley & Sons.)

As shown in Figure 10.10, the truss achieves the 50 ft span with single-piece top chords of timber. Including the projected overhangs, these members are approximately 31 ft long, not an unreasonable length for a timber piece. However, the bottom chord is too long; you can anticipate needing a joint for the bottom chord somewhere near midspan.

The general idea behind the composite wood and steel truss in Figure 10.9 is to use the relatively stout wood members for the truss compression elements and the slim steel members for tension elements. In addition, you limit the number of tension joints between wood members. The layout in Figure 10.9 uses steel tension members for all but the bottom chords. In Figure 10.10 the tension joints are developed with the welded gusset plates.

Figure 10.11 shows the values determined for a unit gravity loading. These values for internal forces in the members can be used by simple multiplication for various specific loads, as determined by optional forms of the roof construction. Based on the form of construction shown in Figure 10.6c and a minimum roof live load of 20 psf, the total loads on the individual roof purlins that are supported by the truss top chords will be approximately 1800 lb live load and 3600 lb dead load.

Figure 10.12 shows wind load values, based on criteria from the 1994 *Uniform Building Code.* The form of loading is based on the roof slope and a minimum horizontal wind pressure of 20 psf at the roof level.

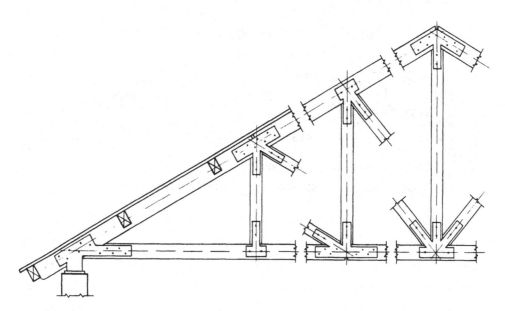

FIGURE 10.10 Form of the timber truss for Building Three, using steel gusset plates and bolts.

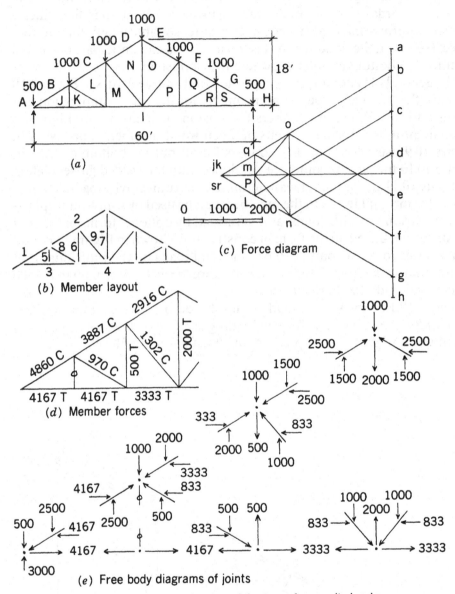

FIGURE 10.11 Investigation of the truss for gravity loads.

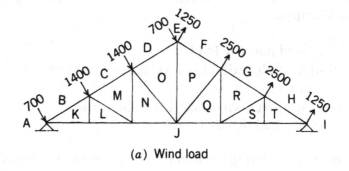

(a) Wind load

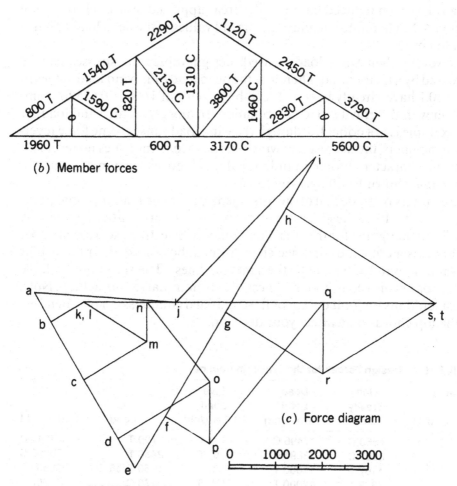

(b) Member forces

(c) Force diagram

0   1000   2000   3000

FIGURE 10.12 Investigation of the truss for wind load.

The results of the investigations in Figures 10.11 and 10.12 are summarized in Table 10.1. When designing you must pay attention to three combinations:

Dead load plus live load.

Dead load plus wind load when the member force is of the same sign (T or C).

Dead load plus wind load when the net is a reversal of the sign of the gravity load member force.

When considering the roof live load, it is usually possible to calculate a stress increase for wood construction. However, the live load in this case already has been reduced based on the area supported by a single truss (see Section A.2). Therefore you must use the full values of the dead load plus live load forces.

However, when wind load is used, design values for stresses can be increased by a factor of 1.6. Therefore the dead load plus wind load values in Table 10.1 have already been reduced by a factor of $1/1.6 = 0.625$ for comparison with the gravity load combination. In any event, the wind load for this example is not critical: although the wind load causes some force reversals in members, the net result of wind plus dead load indicates no reversals. Thus the character of force in individual members as tension or compression is that shown for the gravity loads.

A summary of the design of the truss members for the loads in Table 10.1 is given in Table 10.2. These selections are probably conservative, as you may desire a higher grade for the truss construction. Selecting the truss member thickness is a critical design decision. All members needed for the details shown in Figure 10.10 must be the same thickness. The selections in Table 10.2 are based on a common thickness of 6 in. nominal (5.5 in. actual). Note: Considerations for joint design or the combined compression plus bending on the top chords may affect your decision.

TABLE 10.1 Design Forces for the Truss (in Pounds)

Member (See Figure 10.11)	Unit Gravity Load	Dead Load (1.8 × Unit)	Live Load (3.6 × Unit)	Wind Load	DL + LL
1	4860 C	17496 C	8748 C	3790 T	26244 C
2	3887 C	13994 C	6997 C	2450 T	20990 C
3	4167 T	15000 T	7500 T	1960 T/5600 C	22500 T
4	3333 T	12000 T	6000 T	3170 C	18000 T
5	0	0	0	0	0
6	500 T	1800 T	900 T	820 T/1460 C	2700 T
7	2000 T	7200 T	3600 T	1310 C	10800 T
8	970 C	3492 C	1746 C	1590 C/2830 T	5238 C
9	1302 C	4688 C	2344 C	2130 C/3800 T	7031 C

TABLE 10.2   Selection of the Truss Members

Member (See Figure 10.11)	Member Length (ft)	Design Force (kips)	Member Selection[a]	
			With All 6 In. Nominal Thickness	With All 8 In. Nominal Thickness
1	11.7	26.30 C	6 × 10	6 × 8
2	11.7	21.00 C	6 × 8	6 × 8
3	10.0	22.50 T	6 × 6	6 × 8
4	10.0	18.00 T	6 × 6	6 × 8
5	6.0	0.00	6 × 6	6 × 8
6	12.0	2.70 T	6 × 6	6 × 8
7	18.0	10.80 T	6 × 6	6 × 8
8	11.7	5.24 C	6 × 6	6 × 8
9	15.6	7.03 C	6 × 8	6 × 8

[a] Minimum thickness for code qualification as heavy timber is 6 in. Selections for top chord are without consideration for bending due to purlin loads.

A comparison of the trusses in Figures 10.9 and 10.10 shows a significant difference in the means for achieving the roof overhang. In Figure 10.9 the cantilever is achieved by the rafters, while in Figure 10.10 the top chord of the truss is extended through the support joint.

Other differences deal with how the joints are achieved. In Figure 10.9 the wood-to-wood compression joints are achieved by direct bearing—using member-to-member custom fitting or intermediate bearing blocks. In Figure 10.10 all the joints use steel plate connecting elements. This difference is due to a loss of craft (for achieving the well-fit wood joints) and a gain of welding reliability and technology (for assembling the steel plate connectors).

Of course, you probably won't use the truss in the preceding design unless the historical relationship is important to the overall building design goals. Figure 10.13 shows a slightly more contemporary form for the timber truss, using chords of multiple members with joints achieved by overlapping of members; this joint form permits the use of split-ring connectors. The resulting construction has much smaller joint deformations and so has much better control of deflections.

To use this truss, you must design joint layouts that permit the placing of the required split-rings within the space defined by edge limits, spacing limits, and the sizes of the truss members (that is, their widths).

10.3   MASONRY WALLS

Figures 10.14 and 10.15 show the construction of the masonry walls with CMUs (concrete blocks). The basic modular unit here has nominal dimensions of 16 in. in length, and 8 in. in height.

An alternative construction for the walls is shown in Figure 10.16. The exterior surface is developed with a proprietary EIF system, which is attached to the face of the CMU structural wall. This construction would obviously produce a different exterior appearance; in fact, it would bear little resemblance to any traditional masonry exterior. However, the final finished surface can be altered to simulate many forms of construction. This is the ultimate "Disneyland" finish.

FIGURE 10.13 Alternative form for the wood truss, using multiple-element members.

FIGURE 10.14 Partial elevation of the masonry wall structure.

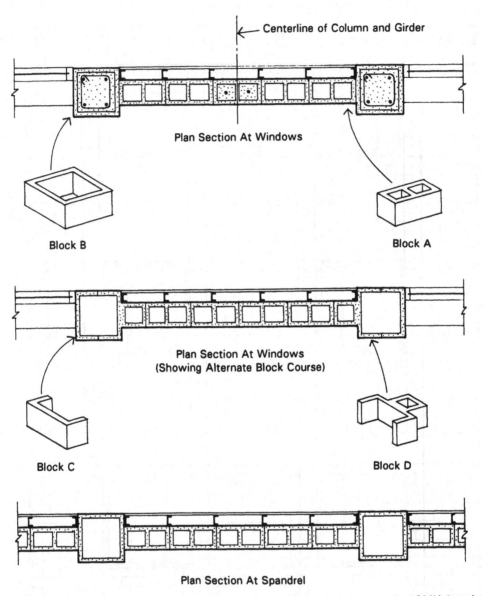

Plan Section At Windows

Block B

Block A

Plan Section At Windows
(Showing Alternate Block Course)

Block C

Block D

Plan Section At Spandrel

FIGURE 10.15 Details of the masonry walls using hollow concrete blocks (CMUs) and reinforced masonry construction.

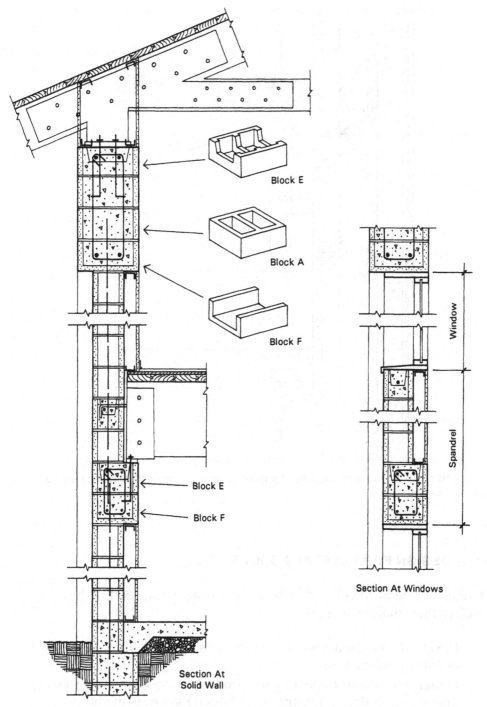

Block E

Block A

Block F

Block E

Block F

Section At
Solid Wall

Window

Spandrel

Section At Windows

FIGURE 10.15 *(Continued)*

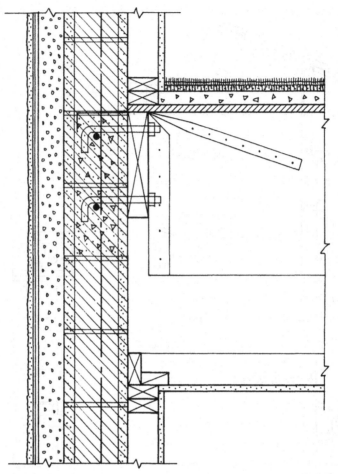

FIGURE 10.16 Alternative construction for the exterior walls, using an exterior insulation finish on the CMU walls.

10.4 DESIGN FOR LATERAL FORCES

Design for lateral forces (either wind or earthquakes) for this building involves three major concerns:

1. Design of the masonry walls as shear walls, with combined lateral and gravity load functions.
2. Design of elements of the roof and floor structures to serve necessary diaphragm, collector, tie, and other lateral load resisting functions.
3. Design of connections between the wood framing and the masonry walls for the necessary transfer of forces due to lateral loads, as well as any forces due to gravity loads.

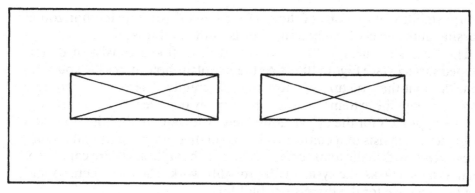

FIGURE 10.17 Schematic plan for the horizontal floor diaphragm.

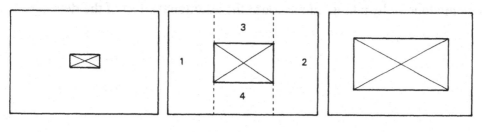

| Small Opening: | Large Hole: | Very Large Hole: |
| functioning of whole diaphragm mainly unaffected; reinforce edges and corners of hole | diaphragm reduced to parts (subdiaphragms) that may work as a connected set | not a diaphragm, can function only as a very stiff rigid frame |

FIGURE 10.18 Range of effects of a hole in a horizontal diaphragm: (a) Small hole with negligible effect on the diaphragm; requires only reinforcement of the hole edges and corners. (b) Large hole, possibly indicating use of subdiaphragms. (c) Hole so large that the diaphragm becomes incapable of effective action; results in essentially no bracing at the diaphragm level.

A complete discussion of how to design this structure for lateral loads (and lateral + gravity load combinations) could take an entire book. Much of the work is illustrated in other examples: determination of loads, investigation for forces in diaphragms, analysis for anchorage requirements, and so on. And Chapter 6 covers investigation of stresses in masonry structures.

What I do not detail elsewhere in this book centers on the problems of the hollowed-out building, as shown in Figure 10.6. The ground-level floor structure and the roof structures are generally continuous surfaces. The roof slope makes it questionable to consider the roof deck as an actual horizontal diaphragm, but most designers disregard the discrepancy as long as the slope does not much exceed that shown here. However, you can improve the lateral stability of the tops of the exterior walls if you provide horizontal trussing at the level of the bottom chords of the roof trusses.

The floor structures at the second and third floors consist of donut-shaped surfaces (see Figure 10.17). As the size of the hole increases, you must consider that the structure will function differently. There is no clear line here, although the extreme cases are reasonably definite (see Figure 10.18).

One approach for the upper floors here is to assume that the floor at a single level consists of a connected set of smaller diaphragms (called *subdiaphragms*) within the total surface. If these subdiaphragms are capable of their assigned tasks, the system will probably work. The loads here should not be excessive for the three-story building.

You also may be able to turn the entire floor structure into a single rigid frame, as is commonly done with vertical planar bents of columns and beams (see discussion in Section 13.3). You can achieve this with the beam system alone or by taking into account the additional effect of the decking.

PART IV

BUILDING FOUR

11

GENERAL CONSIDERATIONS
FOR BUILDING FOUR

11.1 THE BUILDING

Building Four is a modest-sized building, generally characterized as *low rise* (see Figures 11.1 and 11.2). Construction varies widely, although a few popular forms tend to dominate the field.

Some modular planning is usually required, involving the coordination of dimensions for column spacing, window mullions, and interior partitions. This modular coordination may be extended to development of ceiling construction, lighting, ceiling HVAC elements, and the systems for access to electric power, phones, and other signal wiring systems. There is no single magical modular system; all dimensions between 3 and 5 ft have been used. Selecting a particular proprietary system for the curtain wall, interior modular partitioning, or an integrated ceiling system, however, may establish a reference dimension.

For buildings built as investment properties, with speculative occupancies that may vary over the life of the building, it is usually desirable to accommodate future redevelopment of the interior. In other words, you need design with few permanent structural elements: the major structure (columns, floors, and roof); exterior walls; and interior walls enclosing stairs, elevators, rest rooms, and risers for building services. Everything else should be nonstructural or demountable in nature.

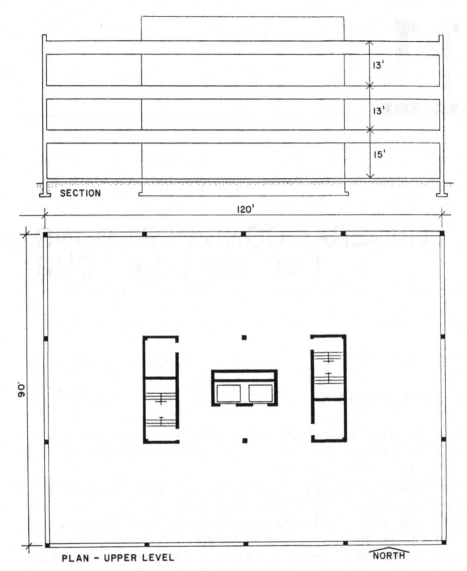

FIGURE 11.1 Building Four: general form.

Interior column spacing should be as wide as possible to reduce the number of freestanding columns in the building plan. In fact, a column-free interior may be possible if the distance from a central core (grouped permanent elements) to the outside walls is not too far for a single span. Column spacing at the building perimeter does not affect remodeling, so additional columns are sometimes used there for better gravity loading or to develop a better perimeter rigid frame system for lateral loads.

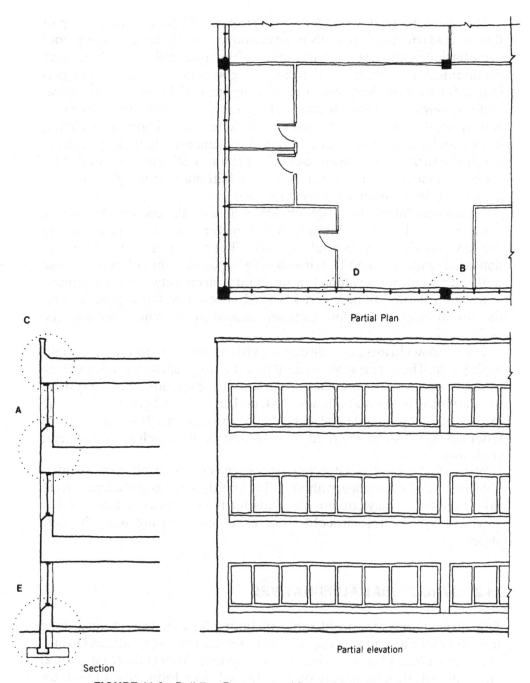

Partial Plan

Partial elevation

Section

FIGURE 11.2 Building Four: general form of the exterior walls.

The space between the underside of suspended ceilings and the top of floor or roof structures typically must contain many elements besides those of the basic construction, including parts of the structural, HVAC, electrical, communication, lighting, and fire-fighting systems. Because depth permitted for the spanning structure and the general level-to-level vertical building height is not easy to changer later, designers must often decide very early in the design process the space required for these elements. Of course, if you provide a generous space for building elements, you make it easier to design the various subsystems. But taller exterior walls and extra height for stairs, elevators, and service risers result in additional cost, so tight control of the level-to-level distance is very important.

A major architectural design choice centers on the construction of the exterior walls. For the column-framed structure, you must integrate the columns and the nonstructural infill wall. The basic form of the construction (see Figures 11.3 and 11.4) involves incorporating the columns into the wall, with windows developed in horizontal strips between the columns. Since the exterior column and spandrel covers develop a general continuous surface, the window units are created as "punched" holes in the wall.

The windows in this example do not exist as parts of a continuous curtain wall system. They are essentially individual units, placed in and supported by the general wall system. The curtain wall is a stud-and-surfacing system, like the typical light wood stud wall system. The studs are light-gage steel, the exterior covering is a system of metal faced sandwich panel units, and the interior covering, where required, is gypsum drywall, attached to the metal studs with screws.

Notice that the wall construction (shown in detail A) results in a considerable interstitial void space. Although filled partly with insulation materials, this space easily may contain elements for the electrical system or other services—for example, in cold climates, a perimeter hot water heating system.

11.2   STRUCTURAL ALTERNATIVES

Structural options are considerable, including the light wood frame (if the total floor area and zoning requirements permit its use) and all steel frame, concrete frame, and masonry bearing wall systems. Your choice of structural elements will depend mostly on the desired plan form, type of window arrangements, and clear spans required for the building interior.

At this building height and taller, the basic structure usually must be either steel or reinforced concrete. Chapter 12 describes using rolled steel shapes for a column-and-beam frame system. Chapter 13 details using sitecast concrete for columns and horizontal beams and slabs. With either of these basic systems, many curtain wall systems are possible.

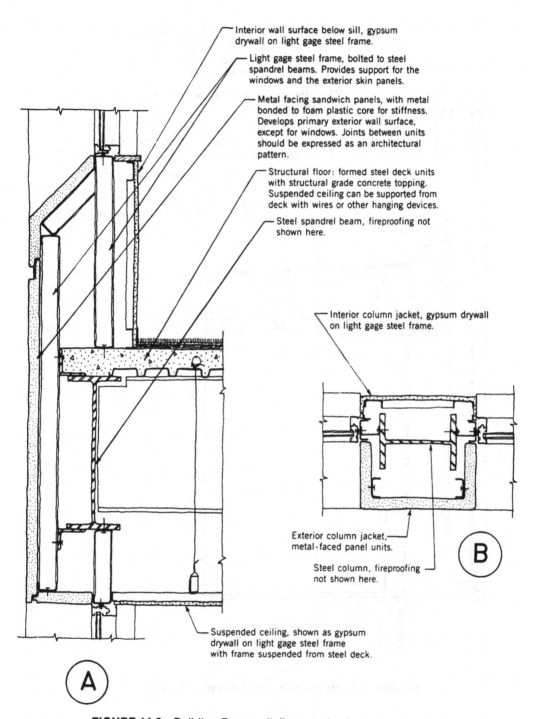

Interior wall surface below sill, gypsum drywall on light gage steel frame.

Light gage steel frame, bolted to steel spandrel beams. Provides support for the windows and the exterior skin panels.

Metal facing sandwich panels, with metal bonded to foam plastic core for stiffness. Develops primary exterior wall surface, except for windows. Joints between units should be expressed as an architectural pattern.

Structural floor: formed steel deck units with structural grade concrete topping. Suspended ceiling can be supported from deck with wires or other hanging devices.

Steel spandrel beam, fireproofing not shown here.

Interior column jacket, gypsum drywall on light gage steel frame.

Exterior column jacket, metal-faced panel units.

Steel column, fireproofing not shown here.

B

Suspended ceiling, shown as gypsum drywall on light gage steel frame with frame suspended from steel deck.

A

FIGURE 11.3 Building Four: wall, floor, and column construction.

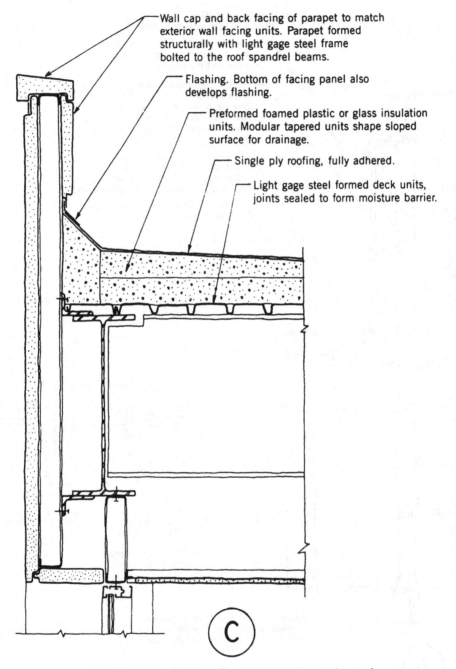

Wall cap and back facing of parapet to match exterior wall facing units. Parapet formed structurally with light gage steel frame bolted to the roof spandrel beams.

Flashing. Bottom of facing panel also develops flashing.

Preformed foamed plastic or glass insulation units. Modular tapered units shape sloped surface for drainage.

Single ply roofing, fully adhered.

Light gage steel formed deck units, joints sealed to form moisture barrier.

C

FIGURE 11.4 Building Four: construction at the roof.

The structural system must take into account both gravity and lateral force problems. Gravity requires developing horizontal spanning systems for the roof and floors and stacking the vertical elements (walls and columns) that provide support for the spanning structure. Meanwhile, the most common choices for the lateral bracing system are the following (see Figure 11.5):

Core Shear Wall System (Figure 11.5*a*). Solid walls produce a very rigid central core. The rest of the structure leans on this rigid interior portion, and the roof and floor constructions outside the core, as well as the exterior walls, are free of concerns for major lateral bracing.

Truss-Braced Core. Similar to the shear-wall- braced core, the solid walls are replaced by bays of trussed framing (in vertical bents), using various possible patterns for the truss elements.

Perimeter Shear Walls (Figure 11.5*b*). Turns the building into a tubelike structure. Because doors and windows must pierce the exterior, the perimeter shear walls usually consist of linked sets of individual walls (sometimes called *piers*).

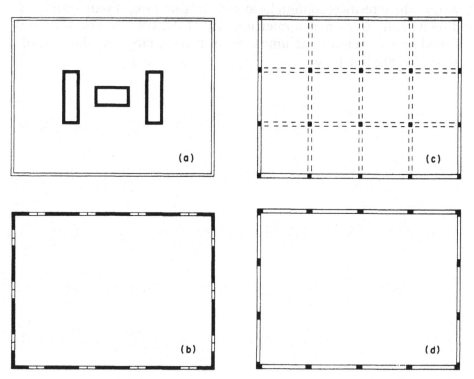

FIGURE 11.5 Building Four: options for development of the lateral bracing system.

Mixed Exterior and Interior Shear Walls. Combines the core and perimeter systems.

Full Rigid-Frame System (Figure 11.5c). Uses the vertical planes of columns and beams in each direction as a series of rigid bents. For this building there would thus be four bents for bracing in one direction and five for bracing in the other direction. This system requires that the beam-to-column connections be moment resistive.

Perimeter Rigid-Frame System (Figure 11.5d). Uses only the columns and beams in the exterior walls, resulting in only two bracing bents in each direction.

In the right circumstances, any of these systems may be acceptable. Each has advantages and disadvantages from both structural design and architectural planning points of view.

The core-braced schemes were once popular, especially when wind was the major concern. The core system allows the greatest freedom in planning the exterior walls. The perimeter system, however, produces the most torsionally stiff building—an advantage for seismic resistance. The rigid-frame schemes permit the free planning of the interior and the greatest openness in the wall planes. The integrity of the bents must be maintained, however, which restricts column locations and planning of stairs, elevators, and duct shafts so as not to interrupt any of the column-line beams. If designed for lateral forces, columns are likely to be large and thus intrude more in the building plan.

12

DESIGN OF THE
STEEL STRUCTURE

Figure 12.1 shows a partial plan for a steel framing system for one of Building Four's upper floors. Shown here is a structure with W-shapes for columns, girders, and beams and a formed sheet steel deck for the floors and roof. This chapter presents the design of this structural system.

12.1 GENERAL CONSIDERATIONS

Because the building has continuous horizontal strip windows, the exterior walls will not accommodate a perimeter shear wall system; as a result, lateral bracing must be achieved with a core bracing system or a development of the steel frame for rigid bent action. If wind is the critical concern, you may select the core bracing. If seismic effects are critical, you should choose the bent system.

Many variations of this basic structural system are possible; most deal with choices for the deck system and the beams that directly support the deck. Some options are as follow:

1. Steel open-web joists for the deck support. These become more feasible as the span is increased. At a 30 ft span, the W sections are competitive; at 40 ft go with the joists. Open-web joists are spaced closer together, which may affect your deck choice.

2. Sitecast concrete slab with steel beams. Includes the development of the beams in composite action with the slab, using lug connectors on top of the beams to engage the cast concrete.
3. Plywood deck with nailable joists (fire codes permitting). This is a presently popular system with the use of either light composite wood and steel trusses (wood chords and steel web members) or the various forms of wood I sections formed with wood flange members and plywood or particleboard webs. Wood 2 × members are bolted to the tops of the steel frame members to permit nailing of the plywood at these locations.

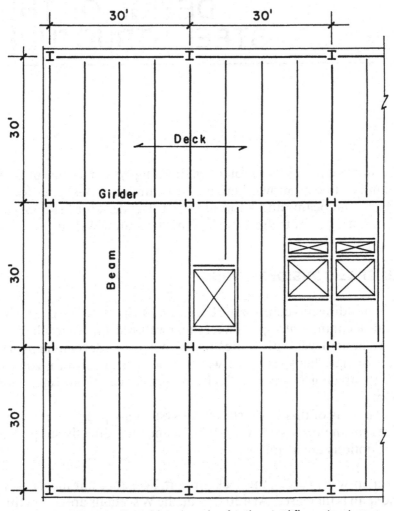

FIGURE 12.1 Partial framing plan for the steel floor structure.

The fact that decking also must develop the horizontal diaphragm action required for wind or seismic effects can affect the choice of decking materials as well as the installation details—in particular, how the deck is attached to the steel frame. The concrete fill on the deck adds considerable weight, so it is sometimes advantageous to achieve the 30% weight savings possible by using a lightweight structural aggregate for the fill.

Figure 12.1 shows a framing system that uses rolled steel beams spaced at some module relating to the column spacing. As shown, the beams are 7.5 ft on center, and the beams that are not on the column lines are supported by column line girders. Thus three-fourths of the beams are supported by the girders and the remainder are supported directly by the columns. The beams in turn support a one-way spanning deck, consisting of formed sheet steel units with a poured-in-place concrete topping.

Within this basic system scheme are a number of variables represented by the elements of the system. These may be considered individually, but they are also inter-related, so that change of one affects the others. In the end, alternate versions of the system must be compared in their complete form; for effectiveness for structural functions, for achievement of architectural purposes, for construction cost, and so on. The principal variable elements of the basic system are:

1. *The Beam Spacing.* This determines the beam loading and thus the size of beam required. However, it also directly affects the choice for a deck, as it defines the deck span. It's actual dimension also relates to the spacing of supports for the beams and girders.

2. *The Deck.* The type and thickness of the deck affects construction dimensions and details. Decks are also typically multi-functional as diaphragms for lateral forces.

3. *Beam/Column Relationship in Plan.* The girders are supported by the columns. They thus define a series of vertical bents with horizontal and vertical members. As shown in Figure 12.1, the beams also start on the column line, perpendicular to the girders. Thus the beams also define a series of vertical bents with the columns. This beam/column relationship may have useful purposes (bracing of columns; producing of lateral rigid frames; and so on), but if none is really required, a simplification in framing can be achieved if the beams do not intersect the columns; permitting use of the same beam-to-girder connection as used for the rest of the beams.

Orientation of the wide-flange steel columns is another consideration. If a rigid frame is used for lateral load resistance, the system shown might be chosen so that there are the same total number of columns in each direction turned to present their major stiffness (as indicated by I_x). This biaxial equalization of total stiffness is somewhat more meaningful for seismic resistance. For wind, the load will be greater in one direction because the building is oblong.

Locations of major openings for duct shafts, elevators, and stairs should be developed so as not to interfere with any of the girders or beams on the column lines. Major plumbing and wiring risers should also be kept off the column lines. These considerations are essential to the integrity of the vertical column-beam bents for the lateral load resisting system.

A similar system would likely be used for the roof of this building. A light steel frame structure is needed for the structure on the roof that covers the elevators, stairs, and rooftop HVAC equipment.

The fireproofing of the steel frame and deck is accomplished as follows:

1. *Top of the steel deck and exposed faces of the spandrel beams and beams at openings:* poured concrete (probably with lightweight aggregate).
2. *Exterior columns, interior sides of beams and girders, and the underside of the steel deck:* sprayed-on fireproofing.
3. *Interior columns:* metal lath and plaster.

A36 steel will be used for all steel frame members.

12.2 THE STEEL FLOOR STRUCTURE

Several options are possible for the floor deck. In addition to structural concerns, which include gravity loading and diaphragm action for lateral loads, you must consider fireproofing and accommodation of wiring and piping for building services. For office buildings you may want a system that incorporates a two-way wiring distribution network for both power and communication in the structural floor.

If the structural floor deck is a concrete slab either precast or poured in place, the wiring network is usually buried completely in a nonstructural concrete fill on top of the structural slab. If a steel deck is used, you may be able to use closed cells of the formed deck for wiring in one direction, reducing the wiring in the fill to that for the perpendicular direction only. The latter system is assumed for this design, using a steel deck with a 1.5 in. depth of the corrugations and a fill of 2.5 in. on top of the deck. Total dead weight of this deck combination depends on the thickness of the sheet steel, the profile of the formed deck units, and the unit density of the concrete fill. For this example, however, assume that the deck weights 35 psf [1.68 kPa], which is an average value with lightweight concrete fill. In addition, assume a dead load of 15 psf [0.72 kPa] for the ceiling, lights, ducts, and supported fixtures. Thus the total dead load superimposed on the deck is 50 psf [2.40 kPa].

In addition to this dead load, *UBC* Section 1604.4 requires a load of 20 psf [0.96 kPa] to account for movable partitions. Although this is technically a dead load, its randomness of occurrence makes it similar to live load. For the typical office areas in office buildings, the specified live load is 50 psf [2.40 kPa]; the *UBC* also stipulates that the floor must sustain a concentrated load of 2000 lb [8.90 kN]. For corridors and lobbies, meanwhile, the code requires a live load of 100 psf [4.79 kPa]. Since it is not necessary to consider partitions in the corridor or lobby space, the comparison is really between 100 psf in some areas and 70 psf in others. This difference presents a problem in buildings designed to accommodate remodeling because no one can predict the exact location of corridors and lobbies. A conservative approach is to design for 100 psf live load for the entire floor, ignoring the partition loading. As a compromise, use the following loadings for the various elements of the structure:

1. For the deck
   Live load = 100 psf
   Dead load = 50 psf, including the deck weight
2. For the beams that directly support the deck
   Live load = 80 psf (100 psf if combined with partitions)
   Dead load = 70 psf, including partitions but not the weight of the beams
3. For girders and columns
   Live load = 50 psf
   Dead load = 70 psf + weights of beams, girders, columns, and walls

Although some generalized information is available for steel decks, it is best to select deck data and determine necessary design data from manufacturers' catalogs. This example uses a deck that is widely available.

The typical beam is a simple span beam supported by ordinary framing connections to the webs of the girders. For the 30 ft span beam on 7.5 ft centers, the loading is:

Beam DL = 7.5(70) = 525 plf + beam weight, say 560 plf

Live-load reduction = 0.08(A − 150) = 0.08(225 − 150) = 6%

Live load = 0.94(7.5)(80) = 564 plf

Total unit load on beam = 1124 plf

Total supported load = 1.124(30) = 33.7 kips [150 kN]

For consideration of bending moment only, Table B.17 yields choices of W 16 × 45, W 18 × 46, or W 21 × 44. Your actual selection may be influenced by the depth of the girders or the need for space beneath the beams for ducts or other items. Figure B.3 shows that the shallowest option beam, the 16-in.-deep section, is not critical for a total load deflection of $L/240$.

Figure 12.2 shows the loading condition for the girder, assuming only the loads due to the supported beams. While this ignores the effect of the uniformly distributed load of the girder weight, it is reasonable for an approximate design, since the girder weight is a minor loading. For the beam load, determine the following:

Total beam DL $= 0.570 \times 30 = 17.1$ kips.

Live-load reduction $= 0.08(A - 150) = 0.08(675 - 150) = 42\%$.

(Note that the girder carries three beams, or three-fourths of one full column bay of 900 ft².) The maximum permitted reduction is 40%; thus

total beam live load $= 0.60(0.050)(7.5)(30) = 6.75$ kips

total load on girder $= 17.1 + 6.75 = 23.85$ say, 24 kips [106 kN]

Selection of a member for this situation may be made using various data sources. Tables and graphs can be used relating selections only to the maximum bending moment; thus requiring no additional computations. However, another simple technique is to compute an equivalent uniform load, using the equation for simple beam moment, thus:

$$\text{Maximum moment: } M = \frac{WL}{8}$$

and, thus

$$\text{Equivalent total uniform load: } W = \frac{8M}{L}$$

for this case

$$W = \frac{(8)(360)}{30} = 96 \text{ kips}$$

This hypothetical load—the equivalent uniform load (EUL) or equivalent tabular load (ETL)—can be used in some tables; Table B.17 being an example. Using the 96 kip load as computed, Table B.17 yields possibilities for several beams, the lightest weight selection being a W 24 × 84.

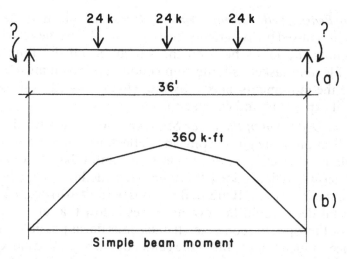

FIGURE 12.2 Approximation of the gravity load moment on the girder.

It is possible, however, that some other section may be desirable—perhaps for its increased stiffness or for the form of its flanges to facilitate the welded connections. For example, from Table B.17, you can see that there is a deeper beam available at the same weight as the W 24 × 84: namely, the W 27 × 84. After comparing the properties for the moments of inertia for these two beams, it may be determined that the 27 in. beam will have approximately 20% less deflection—an improvement obtained at no additional steel costs. However, the additional 3 in. will probably increase the story-to-story height, which becomes a critical issue.

The spandrel framing (the beams and girders at the building edge) has less gravity loads from the floors, but it must carry the building curtain wall. If the wall construction is heavy, the spandrel members may have a total load similar to that of the interior members. When selecting the spandrel members, deflection may be more critical than when selecting interior members, as it involves distortions of the curtain wall construction. On the other hand, depth is less restricted, and a deeper and stiffer member may be possible. As with the concrete structure in Chapter 13, it may be possible to use a very deep spandrel beam to develop a perimeter bent bracing system.

You must also take into account the tolerable vertical deflection of the floor and roof systems. This is a complex issue involving more judgment than facts. Some of the specific considerations are

Bounciness of the Floors. Bounce is essentially a matter of the stiffness and fundamental period of vibration of the deck. Use of static deflection limits generally recommended usually assures reasonable lack of bounce.

Transfer of Bearing to the Curtain Wall and Partitions. The deformations of the frame caused by live gravity loads and wind must be considered in developing the joints between the structure and the nonstructural walls. Flexible gaskets, sliding connections, and so on must be used to permit the movements caused by these loads as well as those due to thermal expansion and contraction.

Dead-Load Deflection of the Floor Structure. As shown in Figure 12.3, the deflection of the girders plus the deflection of the beams creates a cumulative deflection at the center of a column bay. Given the normally permitted deflections, this cumulative deflection can be a problem for the 30 ft spans. If this deflection is permitted to occur, the top of the metal deck would be several inches below that at the columns. Because the top of the concrete fill must be as flat as possible, this could produce a much thicker fill in the center of the bays. You can compensate by specifying a camber for the beams approximately equal to the calculated dead-load deflection.

12.3 COLUMN DESIGN FOR GRAVITY LOADS

With the core bracing system for lateral load resistance, the exterior columns are all basically designed for vertical gravity loads only. With the steel beams having essentially pin connections to the columns (negligible moment transfer capacity), the exterior columns may be designed for axial compression force only.

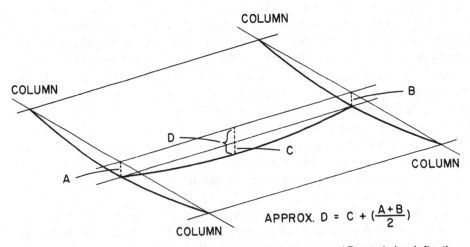

$$\text{APPROX. } D = C + \left(\frac{A+B}{2}\right)$$

FIGURE 12.3 Cumulative deflection of the floor structure: A and B are girder deflections; C is the beam deflection; D is the cumulative deflection.

The interior columns are all involved in core framing conditions (see Figure 12.1), making the determination of their gravity loads somewhat more complicated. Additionally, some of the core columns will be involved in the development of the trussed bents for lateral loads. Figure 12.4 shows a revised core plan with some additional columns that will be used to develop the trussed bents (see Section 12.4).

Figure 12.5 is a common form of tabulation used to determine the column loads. For the exterior columns, there are three separate load determinations, corresponding to the three-story-high columns. For the interior columns, the table assumes that there is a rooftop structure (penthouse) used to house the HVAC and elevator equipment, thus creating a fourth story for these columns.

Figure 12.5 gives loads for the two cases of typical exterior columns. It also offers values for a hypothetical interior column, ignoring the core conditions and assuming a full load periphery of 30 ft × 30 ft, to illustrate the procedure for such a column and also to give some approximation for the size selection of the columns on the major column lines.

Figure 12.5 is organized to facilitate the following determinations:

1. Dead load on the portion of the horizontal structure supported by the column; calculated as an estimated unit load per square foot times the load support periphery. Loads developed in the process of design of the horizontal framing may be used to estimate the unit loads.

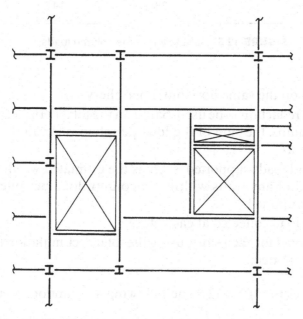

FIGURE 12.4 Modified framing plan for development of the trussed core bents.

Level	Load Source	Corner Column 225 ft²			Intermediate Exterior Column 450 ft²			Interior Column 900 ft²		
		DL	LL	Total	DL	LL	Total	DL	LL	Total
P'hse.	Roof							8	5	
Roof	Wall							5		
	Total/level							13	5	
	Design load									18
Roof	Roof	9	5		18	9		36	18	
	Wall	10			10			10		
	Column	3			3			3		
	Total/level	22	5		31	9		49	23	
	Design load			27			40			72
3rd	Floor	16	11		32	23		63	45	
Floor	Wall	10			10			10		
	Column	3			3			3		
	Total/level	51	16		76	32		125	68	
	LL reduction	24%	12		60%	13		60%	27	
	Design load			63			89			152
2nd	Floor	16	11		32	23		63	45	
Floor	Wall	11			11			11		
	Column	4			4			4		
	Total/level	82	27		123	55		203	113	
	LL reduction	42%	16		60%	22		60%	45	
	Design load			98			145			248

FIGURE 12.5 Tabulation of the column loads.

2. Live load on the same horizontal periphery.
3. Live load reduction to be used, based on the total periphery; for lower-story columns, the sum of the load peripheries for all the levels that they support.
4. Other dead loads supported, such as the estimated weight of the columns and of any walls within the column load periphery that are directly supported.
5. The total load collected at each story.
6. A design load for each story, using the total accumulation from all the stories supported.

For the entries in Figure 12.5, the following assumptions were made:

Roof unit live load = 20 psf (reducible).
Roof dead load = 40 psf. (Estimated, based on the floor construction.)

Penthouse floor live load = 100 psf (for equipment).

Penthouse floor dead load = 50 psf.

Floor live load = 50 psf (reducible).

Floor dead load = 70 psf (including partitions).

Interior walls weigh 15 psf/ft^2 of wall surface area.

Exterior walls average 25 psf/ft^2 of wall surface area.

Figure 12.6 summarizes the column designs for the three columns for which load tabulations are presented in Figure 12.5. For the pin-connected frame, a *K* factor of 1.0 is assumed for the columns, and the full-story heights are used as the unbraced column lengths.

Although column loads in the upper stories are quite low, and some small column sizes would be adequate for the loads, a minimum size of 10 in. is used for the W-shaped columns for two reasons.

The first involves the form of the horizontal framing members (W shapes, in this case) and the type of connections between the columns and the horizontal framing. Regardless of their orientation in plan, the H-shaped columns must usually facilitate the connection of framing on both of their axes. If ordinary beam framing is used, employing connection angles and field erection bolting, a minimum size column is required for practical installation of the angles and bolts.

Level	Story	Unbraced Height (ft)	Corner Column		Intermediate Exterior Column		Interior Column	
			Design Load (kips)	Column Choices	Design Load (kips)	Column Choices	Design Load (kips)	Column Choices
Roof								
	3rd	13	27	W10X33	40	W10X33	72	W10X39
3rd Floor								
	2nd	13	63	W10X33	89	W10X33	152	W10X39
2nd Floor			Assumed location of column splice					
	1st	15	98	W10X33	145	W10X33	248	W10X49
1st Floor								

FIGURE 12.6 Summary of the column design.

Figure 12.7 shows the plan layout of a column and horizontal framing at the building corner. For framing members attached to the column flanges, a minimum width of column flange is required—usually at least 6 in. For framing members attached to the column webs, a minimum column depth is required in order to have a sufficient distance of width for the flat portion of the column web to which the angle legs can be attached. The sizes of bolts required, the minimum angle leg width to accommodate the bolts, and the actual thickness of the beam web will determine this required dimension for the column in a specific case.

A second reason is the problem of achieving splices in the column, where the building height makes a single-piece column impractical. The length of a single column piece that can be handled for transportation and erection depends mostly on the size of the member. A 6-in. W shape will become as flexible as a noodle at a length of 40 ft. A 14-in. W shape, on the other hand, can be handled at some considerable length. The length issue and the chosen column size will generally determine whether a splice is necessary.

If the column can be produced as a single piece (in our case, slightly longer than 40 ft), the splice need not be considered. However, if a splice is required, it is generally most easily achieved with two members of the same nominal depth.

For this design example, assume that a 10 in. W shape is the minimum desirable size to comfortably accommodate the framing. In addition, anticipate using a splice—that is, any size changes should be made within the range of available 10 in. W shapes. If the steel erector decides that a splice is

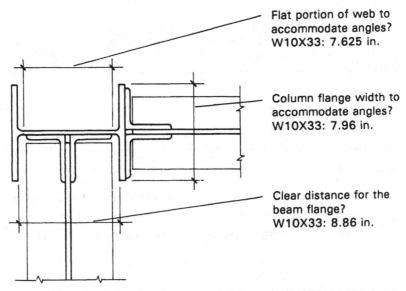

Flat portion of web to accommodate angles? W10X33: 7.625 in.

Column flange width to accommodate angles? W10X33: 7.96 in.

Clear distance for the beam flange? W10X33: 8.86 in.

FIGURE 12.7 Accommodation of beam framing at the columns.

not necessary, the cost of providing a larger column for upper stories can be offset by the elimination of the splice cost.

With all these considerations, the trial set of column sizes is given in Figure 12.6. From sizes indicated in Table B.18, a W 10 × 33 is selected as the minimum size column, providing a flange width of approximately 8 in. For the final selections, it is assumed that a splice occurs just above the second-floor level. It should be noted that Table B.18 presents only a limited number of the available W shapes; a full tabulation is in the *AISC Manual* (Ref. 2).

12.4 DESIGN FOR WIND

Note that in Figure 12.4 there are some extra columns in the framing plan at the location of the stair, rest room, and elevator walls. These columns are used in conjunction with the regular columns and some of the horizontal framing to define vertical planes of framing for the development of the truss bracing system shown in Figure 12.8. These frames will be braced for resistance to lateral loads by the addition of diagonal X-braces. For a simplified design, consider the diagonal members to function only in tension, making the vertical frames consist of statically determinate trusses. There are thus four vertically cantilevered trusses in each direction, placed symmetrically at the building core.

With the symmetrical building exterior form and the symmetrically placed core bracing, this is a reasonable system to use in conjunction with the horizontal roof and floor structures to develop resistance to horizontal forces due to wind or seismic actions. I will illustrate the design of the trussed bents for wind.

The 1994 edition of the *Uniform Building Code* (Ref. 1) provides for the use of the projected profile method for wind using a pressure on vertical surfaces defined as

$$p = C_e C_q q_s I$$

where C_e includes concerns for the height above grade, exposure conditions, and gust effects.

and C_q is the pressure coefficient, which the UBC defines as follows:

$$C_q = 1.3 \text{ for surfaces up to 40 ft above grade}$$
$$= 1.4 \text{ from 40 ft up}$$

The symbol q_s stands for the wind stagnation pressure as related to wind speed and measured at the standard height above ground of 10 m (approx-

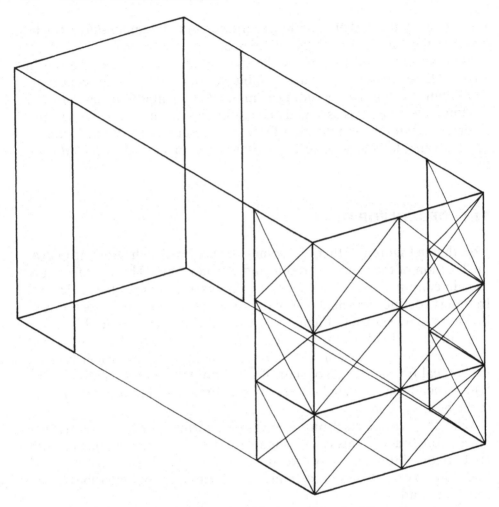

FIGURE 12.8 Development of the trussed bents at the core.

imately 30 ft). For the wind speed of 80 mph assumed earlier, the UBC yields a value for q_s of 17 psf.

Table 12.1 summarizes the foregoing data for the determination of the wind pressures at the various height zones on Building Four. For investigation of the wind effects on the lateral bracing system, the wind pressures on the exterior wall are translated into edge loadings for the horizontal roof and floor diaphragms, as shown in Figure 12.9. Note that we have rounded off the wind pressures from Table 12.1 for use in Figure 12.9.

The accumulated forces noted as H_1, H_2, and H_3 in Figure 12.9 are shown applied to one of the vertical trussed bents in Figure 12.10a. For the east–west bents, the loads will be as shown in Figure 12.10b. These loads are determined by multiplying the edge loadings for the diaphragms as shown in

TABLE 12.1 Design Wind Pressure for Building Four (Exposure Condition B)

Height Above Ground (ft)	C_e	C_q	Pressure[a] p (psf)
0–15	0.62	1.3	13.7
15–20	0.67	1.3	14.8
20–25	0.72	1.3	15.9
25–30	0.76	1.3	16.8
30–40	0.84	1.3	18.6
40–60	0.95	1.4	22.6

[a] Horizontally directed pressure on vertical surface: $p = C_e \times C_q \times 17$ psf.

Figure 12.9 by the 92 ft overall width of the building on the east and west sides. For a single bent, this total force is divided by four. Thus for H_1 the bent load is determined as

$$H_1 = 195 \times 92 \div 4 = 4485 \text{ lb}$$

Analyzed as a truss, ignoring the compression diagonals, the resulting internal forces in the bent are as shown in Figure 12.10c. The forces in the diagonals may be used to design tension members, using the usual one-third

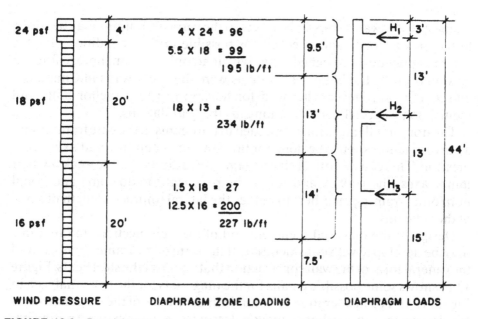

FIGURE 12.9 Development of wind loads, as delivered to the upper level horizontal diaphragms.

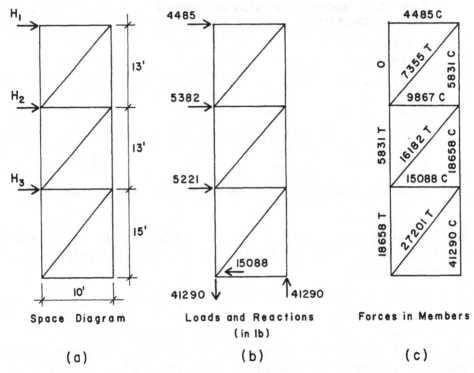

FIGURE 12.10   Analysis of a single trussed bent.

increase in allowable stress. The forces in the vertical columns may be added to the gravity loads and checked for possible critical conditions for the columns previously designed for gravity load only. The anchorage force in tension (uplift) should be dealt with as were the shear walls in Chapters 5 and 9. This may reveal the need for tension-resistive anchor bolts and require some special considerations for the foundations.

The horizontal forces must be added to the beams in the core framing and an investigation should be done for the combined bending and axial compression. This can be critical since beams are ordinarily quite weak on their minor axes (the y-axis), and it may be practical to add some additional horizontal framing members to reduce the lateral unbraced length of some of these beams.

Design of the diagonal members and of their connections to the frame must be developed with consideration of the form of the frame members and the general form of the wall construction that encloses the steel bents. Figure 12.11 shows some possible details for the diagonals for the bent analyzed in Figure 12.10. A consideration to be made in the choice of the diagonal members is the necessity for the two diagonals to pass each other in the midheight of a bent level. If the most common truss members—double angles—are used, it will be necessary to use a joint at this crossing, and the added details for the bent are considerably increased.

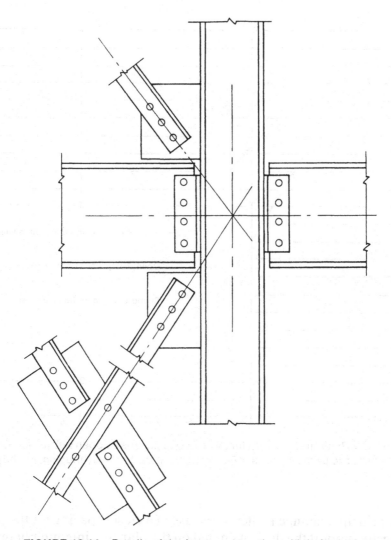

FIGURE 12.11 Details of the bent construction with bolted joints.

An alternative to the double angles is to use either single angles or channel shapes. These may cross each other with their flat sides back-to-back without any connection between the diagonals. However, this involves some degree of eccentricity in the member loadings and the connections, so they should be designed conservatively. It should also be carefully noted that the use of single members results in single shear loading on the bolted connections.

12.5 ALTERNATIVE CONSTRUCTION WITH STEEL TRUSSES

A framing plan for the typical upper floor of Building Four is shown in Figure 12.12. By comparison with the plan in Figure 12.1, it may be observed

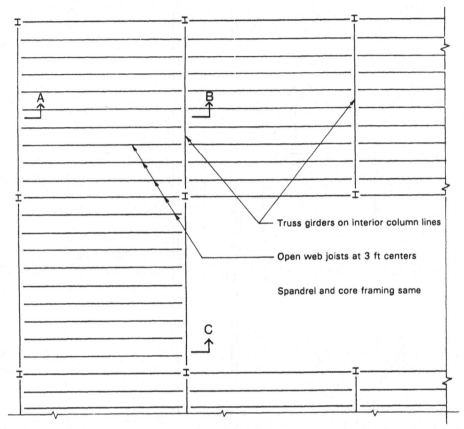

FIGURE 12.12 Partial framing plan for the floor structure using open web joists. Framing for core assumed to be the same as in Figure 12.1. Lateral bracing (bridging) for the joists not shown.

that the principal change relates to the use of a closer spacing for the joists that directly support the deck. As noted on the plan, the intention here is to use open-web steel joists for the joists and steel joist girders for the members that support the joists on the interior.

There are various possibilities for development of the spandrels at the building edge and the core framing for this system. In Figure 12.12 it is assumed that the spandrels and the core framing are essentially the same as in the system using all W-shaped elements for the frame.

There are also various alternatives for development of a lateral bracing system with this scheme. One possibility, however, is the use of the same core trussed bents as in Section 12.4.

Although considerable use is thus made of the same W- shaped framing for the core, approximately 80% of the floor and roof surfaces would be supported by the alternative structure in this scheme. Although somewhat more applicable to longer spans and lighter loads (roofs mainly), this system is frequently used for the situation described here.

One potential advantage of using the truss-form joists and girders is the degree of open nature of the interstitial space between the ceilings and the floor or roof above. This space can be used with much greater freedom for the necessary ducts, piping, and wiring than one generated with solid web beams.

Design of the Joists

The open-web steel joists may be designed by the same basic process as that used for the W-shaped joists in Section 12.2. These, as well as the joist girders, will be provided as proprietary products by some manufacturer, which may also contract or franchise the erection of the system. Determination of specific members will be made from the catalogs of the standard products of such a supplier, which will also supply considerable information for suggested construction details and construction specifications.

Using the data from this case, a joist for this situation is designed in Figure 12.13. This design produces an answer reflecting the lightest member obtainable from Table B.16. While this represents some efficiency for economy, a major concern with these light structures is often that of dynamic deflection (bounciness) when they are used for floors. I therefore recommend consideration of the use of the deepest feasible member for the situation without excessive increase in steel weight or compromise of the interstitial space. In general, increasing the depth or height of the trusses will produce reductions in both static deflection (sag) and dynamic deflection (bounce).

Design of the Joist Girders

The joist girders will also be supplied by the supplier of the joists. Again, although there are industry standards for design and general form, the specific products of the individual supplier must be determined. The pattern of the truss members is relatively fixed and relates to the spacing of the joists. To achieve a reasonable proportion for the panel units of the trusses, the dimension for the depth of the truss girders should be approximately the same as that of the joist spacing. This relates to the achievement of the total dimension for the interstitial space (from underside of the ceiling to top of the finish floor) and to the desired story height (floor-to-floor).

The design of a truss girder for this situation is illustrated in Figure 12.14. The assumed depth of the girder is 3.0 ft, which should probably be considered as a *minimum* depth for this structure. Any additional height that can be obtained will probably reduce the amount of steel used and reduce deflections.

Computations for the Open Web Steel Joists

Joists at 3′ centers, span of 30′ (+ or -)

Loads:

DL = (70 psf X 3′) = 210 lb/ft (not including weight of joist)

LL = (100 psf X 3′) = 300 lb/ft (not reduced)
This is a high load for the offices, but allows a corridor anywhere. Also helps reduce deflection to eliminate floor bounciness.

Total load = 210 + 300 = 510 lb/ft + joist weight

Selection of Joist from Table B.16:

Note that the table value for total load allowable includes the weight of the joist, which must be subtracted to obtain the load the joist can carry in addition to its weight.

Table data: total load = 510 lb/ft + joist, LL = 300 lb/ft, span = 30′

Options from Table B.16:

24K9, total allowable load = 544 - 12 = 532 lb/ft, OK

26K9, stronger than 24K9, only 0.2 lb/ft heavier

28K8, stronger, only 0.7 lb/ft heavier

30K7, stronger, only 0.3 lb/ft heavier

Note that the loading results in a shear limit for the joists, rather than a bending stress limit.

All of the joists listed are economically equivalent, so choice would probably be made with regard to dimensional considerations for the total depth of the floor construction. A shallower joist means a shorter story and less overall building height. A deeper joist gives more space in the floor construction for accommodation of ducts, etc. and probably the least floor bounce.

FIGURE 12.13   Summary of design for the open web joist.

Construction Details for the Truss Structure

Figure 12.15 shows some details for the construction of the trussed system. The deck shown here is essentially the same as that in the scheme with the W-shaped framing. It may be possible to use a lighter-gauge deck because of the reduced span in this scheme. However, the design of the decks for horizontal diaphragm actions must also be considered.

For the joist girder:

Use 40% LL reduction with LL of 50 psf.
Then, LL = 0.60(50) = 30 psf, and the total joist load is

(30 psf)(3 ft c/c)(30 ft span) = 2700 lb, or 2.7 kips

For DL, add partitions of 20 psf to other DL of 40 psf

(60 psf)(3 ft c/c)(30 ft) = 5400 lb

+ joist weight at (10 lb/ft)(30 ft) = 300 lb

Total DL = 5400 + 300 = 5700 lb, or 5.7 kips

Total load of one joist on the grider:

DL + LL = 2.7 + 5.7 = 8.4 kips

Girder specification for choice from manufacturer's load tables:

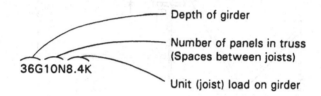

36G10N8.4K
— Depth of girder
— Number of panels in truss
(Spaces between joists)
— Unit (joist) load on girder

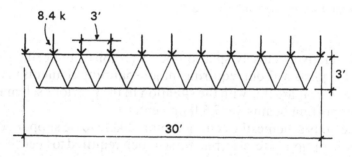

FIGURE 12.14 Summary of design for the joist girder.

There is a slight increase in the overall height of the structure in this system due to the fact that the joist ends sit on top of their supports. The W-shaped joists, though, have their tops level with the supporting girders. This typically adds only 2.5 in., but this is a carefully monitored dimension in design of multistory construction.

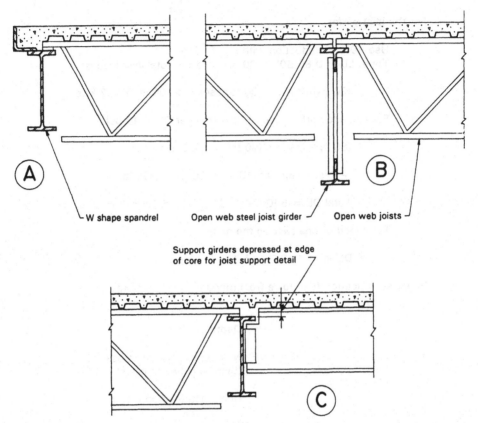

FIGURE 12.15 Details for the floor system with open web joists and joist girders. For locations see the framing plan in Figure 12.12.

With the relatively closely spaced open-web joists, the ceiling may be hung from the joists without requiring major spanning elements in the ceiling structure. Compare this with the situation in the previous scheme where the deck-supporting beams are 7.5 ft on center.

In the previous scheme, the ceiling is more likely to be supported by the deck; this is an option here, also, but not so much required to keep the ceiling structure light.

13

DESIGN OF THE CONCRETE STRUCTURE

In this chapter I present an alternative structure for Building Four: a sitecast, reinforced concrete frame.

The various construction details covered here relate to the partially exposed concrete structure, but it is possible to use a concrete frame with the general exterior wall system shown in Chapter 12.

The concrete structure designed here could be exposed on both the exterior and interior because the basic construction is resistant to weather (exterior) and fire (interior).

A structural framing plan for the upper floors in Building Four is presented in Figure 13.1, where the use of a poured-in-place slab and beam system of reinforced concrete is indicated. Support for the spanning structure is provided by concrete columns. The system, for lateral load resistance is that shown in Figure 11.5d, which utilizes the exterior columns and spandrel beams as rigid-frame bents. This is a highly indeterminate structure for both gravity and lateral force design, and its precise engineering design would undoubtedly be done with a computer-aided system. I discuss the major design considerations and illustrate the use of some simplified techniques for an approximate analysis and design of the structure.

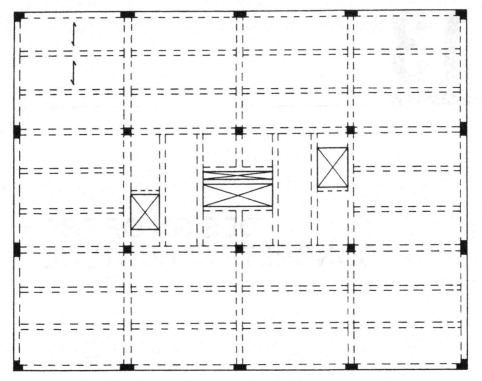

FIGURE 13.1   Framing plan for the concrete floor structure.

13.1   THE CONCRETE FLOOR STRUCTURE

As shown in Figure 13.1, the basic floor-framing system consists of a series of beams at 10-ft centers that support a continuous, one-way spanning slab and are supported by column-line girders or directly by the columns. I discuss the design of three elements of this system: the continuous slab, the four-span beam, and the three-span spandrel girder.

The design conditions for slab, beam, and girder are indicated in Figure 13.2. Shown on the diagrams are the positive and negative moment coefficients derived from those as given in Chapter 8 of the ACI Code (Ref 3). Use of these coefficients is quite reasonable for the design of the slab and beam. For the girder, however, the presence of the concentrated loads makes the use of the coefficients improper according to the ACI Code. But for an approximate design of the girder, the use will produce some reasonable results.

Figure 13.3 shows a section of the exterior wall that demonstrates the general nature of the construction. The exterior columns and spandrel beams are exposed to view and would receive some special treatment for a higher degree of control of the finished concrete. The use of the full available depth of the spandrel beams results in a much stiffened frame on the building exterior, which partly justifies the choice of the peripheral bent system for lateral bracing.

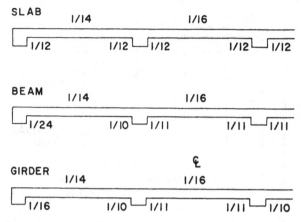

FIGURE 13.2 Approximation factors for bending moments in the slab and beam floor structure.

The design of the continuous slab is presented in Figure 13.4. The use of the 5 in. slab is based on assumed minimum requirements for fire protection. If a thinner slab is possible, the 9 ft clear span would not require this thickness based on limiting bending or shear conditions or recommendations for deflection control. If the 5 in. slab is used, however, the result will tend to be a slab with a relatively low percentage of steel bar weight per sq ft—a situation usually resulting in lower cost for the structure.

The unit loads used for the slab design are determined as follows:

Floor Live Load
    100 psf (at the corridor) [4.79 kPa]
Floor Dead Load
    Carpet and pad at 5 psf
    Ceiling, lights, and ducts at 15 psf
    2-in. lightweight concrete fill at 18 psf
    Assumed 5-in. thick slab at 62 psf
        Total dead load: 100 psf [4.79 kPa]

Inspection of the framing plan in Figure 13.1 reveals that there are a large number of different beams in the structure for the floor with regard to individual loadings and span conditions. Two general types are the beams that carry only uniformly distributed loads as opposed to those that also provide some support for other beams; the latter produce a load condition consisting of a combination of concentrated and distributed loading. Now consider the design of one of the uniformly loaded beams.

The beam that occurs most often in the plan is the one that carries a 10 ft-wide strip of the slab as a uniformly distributed loading, spanning between columns or supporting beams that are 30 ft on center. Assuming the sup-

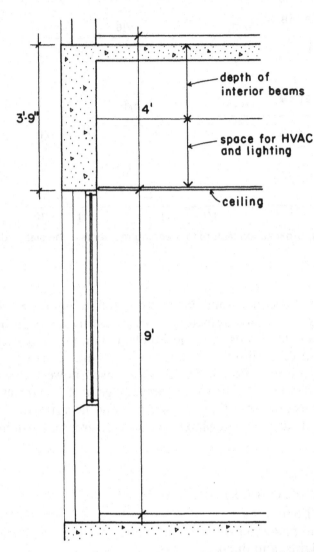

FIGURE 13.3 Form of the exterior wall construction assumed for the concrete structure.

ports to be approximately 12 in. wide, the beam has a clear span of 29 ft and a total load periphery of 29 × 10 = 290 ft². Using the UBC provisions for reduction of live load,

$$R = 0.08(A - 150)$$

$$= 0.08(290 - 150) = 11.2\%$$

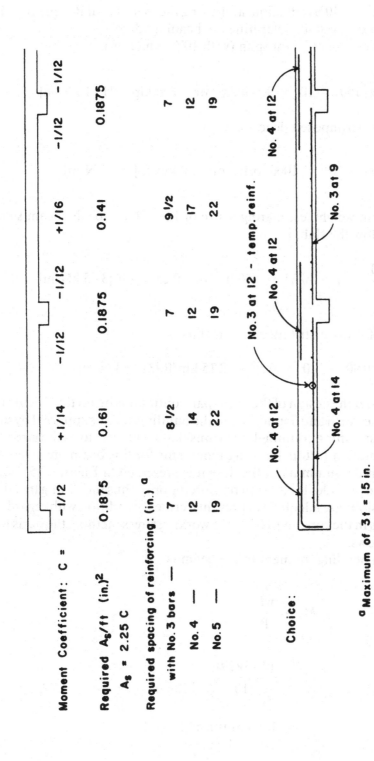

Moment Coefficient: C =

	−1/12	+1/14	−1/12	−1/12	+1/16	−1/12	−1/12

Required A_s/ft (in.)^2

	0.1875	0.161	0.1875	0.1875	0.141	0.1875

$A_s = 2.25\ C$

Required spacing of reinforcing: (in.) [a]

with No. 3 bars —	7	8½	7	7	9½	7
No. 4 —	12	14	12	12	17	12
No. 5 —	19	22	19	19	22	19

Choice:

No. 4 at 12 No. 4 at 14 No. 4 at 12 No. 3 at 12 - temp. reinf. No. 3 at 9 No. 4 at 12

[a] Maximum of 3t = 15 in.

FIGURE 13.4 Summary of the slab design.

Round this off to a 10% reduction, and, using the loads tabulated previously for the design of the slab, determine the beam loading.

Live load per foot of beam span (with 10% reduction):

$$0.90 \times 100 \times 10 = 900 \text{ lb/ft} \quad \text{or} \quad 0.90 \text{ kip/ft } [13.1 \text{ kN/m}]$$

Slab and superimposed dead load:

$$100 \times 10 = 1000 \text{ lb/ft} \quad \text{or} \quad 1.0 \text{ kip/ft } [14.6 \text{ kN/m}]$$

The beam stem weight, estimating a size of 12×20 in. for the beam stem extending below the slab, is

$$\frac{12 \times 20}{144} \times 150 \text{ lb/ft}^3 = 250 \text{ lb} \quad \text{or} \quad 0.25 \text{ kip/ft } [3.65 \text{ kN/m}]$$

The total uniformly distributed load is thus

$$0.90 + 1.0 + 0.25 = 2.15 \text{ kip/ft } [31.35 \text{ kN/m}]$$

Now consider the design of the four-span continuous beam that occurs in the bays on the north and south sides of the building and is supported by the north–south spanning column–line beams that I will refer to as the girders. The approximation factors for design moments for this beam are given in Figure 13.2, and a summary of the design is presented in Figure 13.5. Note that the design provides for tension reinforcing only, thus indicating that the beam dimensions are adequate to prevent a critical condition with regard to flexural stress in the concrete. Using the working stress method, the basis for this is as follows.

Maximum bending moment in the beam is

$$M = \frac{wL^2}{10}$$

$$M = \frac{(2.15)(29)^2}{10}$$

$$= 181 \text{ kip-ft } [245 \text{ kN-m}]$$

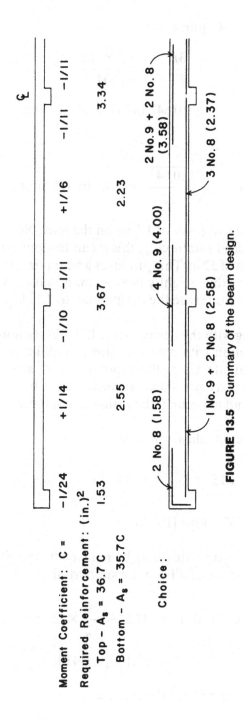

Moment Coefficient: C =	-1/24	+1/14	-1/10	-1/11	+1/16	-1/11	-1/11
Required Reinforcement: (in.)²							
Top - A_s = 36.7 C	1.53		3.67			3.34	
Bottom - A_s = 35.7 C		2.55			2.23		
Choice:	2 No. 8 (1.58)	1 No. 9 + 2 No. 8 (2.58)	4 No. 9 (4.00)		2 No. 9 + 2 No. 8 (3.58)	3 No. 8 (2.37)	

FIGURE 13.5 Summary of the beam design.

Then, for a balanced section,

$$\text{Required } bd^2$$

$$= \frac{M}{R} = \frac{181 \times 12}{0.204}$$

$$= 10{,}647 \text{ in.}^3 \; [175 \times 10^6 \text{ mm}^3]$$

If $b = 12$ in.,

$$d = \sqrt{\frac{10{,}647}{12}} = 29.8 \text{ in. } [757 \text{ mm}]$$

With minimum concrete cover of 1.5 in. on the bars, No. 3 U-stirrups, and moderate-sized flexural reinforcing, this d can be approximately attained with an overall depth of 32 in. This produces a beam stem that extends 27 in. below the slab and is thus slightly heavier than that assumed previously. Based on this size, we increase the design load to 2.25 kip/ft for the subsequent work.

Before proceeding with the design of the flexural reinforcing, it is best to investigate the situation with regard to shear to make sure that the beam dimensions are adequate. Using the approximations given in Chapter 8 of the ACI Code, the maximum shear is considered to be 15% more than the simple span shear and to occur at the inside end of the exterior spans. Consider the following.

The maximum design shear force is

$$V = 1.15 \times \frac{wL}{2} = 1.15 \times \frac{2.25 \times 29}{2}$$

$$= 37.5 \text{ kips } [167 \text{ kN}]$$

For the critical shear stress this may be reduced by the shear between the support and the distance of d from the support; thus

$$\text{Critical } V = 37.5 - \frac{29}{12} \times 2.25$$

$$= 32.1 \text{ kips } [143 \text{ kN}]$$

Using a d of 29 in., the critical shear stress is

$$v = \frac{V}{bd} = \frac{32{,}100}{29 \times 12} = 92 \text{ psi } [634 \text{ kPa}]$$

With the concrete strength of 3000 psi, this results in an excess shear stress of 32 psi that must be accounted for by the stirrups. The closest stirrup spacing would thus be

$$ s = \frac{A_v f_s}{v'b} = \frac{0.22 \times 24{,}000}{32 \times 12} $$

$$ = 13.75 \text{ in. } [348 \text{ mm}] $$

Because this results in quite a modest amount of shear reinforcing, the section may be considered to be adequate.

For the approximate design shown in Figure 13.5, the required area of tension reinforcing at each section is determined as

$$ A_s = \frac{M}{f_s jd} = \frac{C \times 2.25 \times (29)^2 \times 12}{24 \times 0.89 \times 29} $$

$$ = 36.7C $$

Based on the various assumptions and the computations we assume the beam section to be as shown in Figure 13.6. For the beams, the flexural reinforcing in the top required at the supports must either pass over or under the bars in the tops of the girders. Because the girders will carry heavier loadings, it is probably wise to give the girder bars the favored position (nearer the outside for greater value of d) and thus to assume the positions as indicated in Figure 13.6.

At the beam midspans the maximum positive moments will be resisted by the combined beam and slab section acting as a T-section. For this condition, assume an approximate internal moment arm of $d - t/2$ and approximate the required steel areas as

$$ A_s = \frac{M}{f_s(d - t/2)} $$

$$ A_s = \frac{C \times 2.25 \times (29)^2 \times 12}{24 \times (29 - 2.5)} = 35.7C $$

The beams that occur on the column lines are involved in the lateral force resistance actions and are discussed in Section 13.3.

Inspection of the framing plan in Figure 13.1 reveals that the girders on the north–south column lines carry the ends of the beams as concentrated loads at the third points of the girder spans. Consider the spandrel girder that occurs at the east and west sides of the building. This member carries the outer ends of the first beams in the four-span rows and in addition carries a

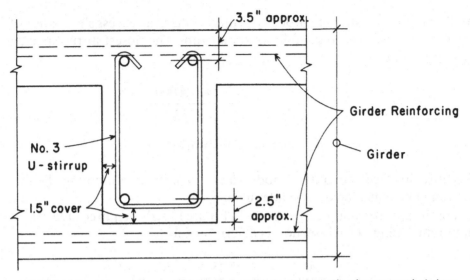

3.5" approx.

Girder Reinforcing

Girder

No. 3
U - stirrup

1.5" cover

2.5"
approx.

FIGURE 13.6 Arrangement of reinforcement in the intersecting beams and girders.

uniformly distributed load consisting of its own weight and that of the supported exterior wall. The form of the girder and the wall was shown in Figure
13.3. From the framing plan, note that the exterior columns are widened in
the plane of the wall to develop the perimeter bent system.

For the spandrel girder, determine the following:

Assumed clear span: 28 ft [8.53 m].

Floor load periphery, based on the carrying of two beams and half the
beam span load, is

$$15 \times 20 = 300 \text{ ft}^2 [27.9 \text{ m}^2]$$

Note: This is approximately the same total load area as that carried by a
single beam, so use the live load reduction of 10% as determined for the
beam.

Loading from the beams:
Dead load: 1.35 kip/ft × 15 ft
= 20.35 kips
Live load: 0.90 kip/ft × 15 ft
= 13.50 kips

Total   33.85 kips, say 34 kips [151 kN]
Uniformly distributed load:

Spandrel beam weight: $\dfrac{12 \times 45}{144 \times 150}$

= 560 lb/ft

Wall assumed at 25 psf: 25×9

$\underline{= 225 \text{ lb/ft}}$

Total

$= 785 \text{ lb/ft},$ say 0.8 kip/ft [11.7 kN/m]

For the uniformly distributed load approximate design moments may be found using the moment coefficients as was done for the slab and beam. Values for this procedure are given in Figure 13.2. The ACI Code does not permit the use of this procedure for concentrated loads, but you may adapt some values for an approximate design using moments for a beam with the third-point loading. Values of positive and negative moments for the third-point loading may be obtained from various references, including Refs. 2, 5, and 6.

Figure 13.7 presents a summary of the work for determining the design moments for the spandrel girder under gravity loading. Moment values are determined separately for the two types of load and then added for the total design moment.

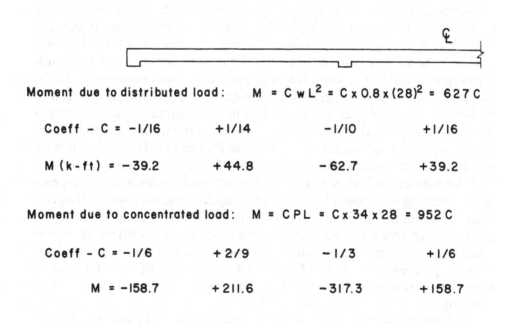

Moment due to distributed load: $M = C w L^2 = C \times 0.8 \times (28)^2 = 627 C$

Coeff – C = –1/16   +1/14   –1/10   +1/16

M (k-ft) = – 39.2   + 44.8   – 62.7   + 39.2

Moment due to concentrated load: $M = C P L = C \times 34 \times 28 = 952 C$

Coeff – C = –1/6   + 2/9   – 1/3   +1/6

M = –158.7   + 211.6   – 317.3   + 158.7

Total gravity – induced moment:

M = –197.9   + 256.4   – 380   + 197.9

FIGURE 13.7 Approximation of the bending moments due to the combined loading on the girder; gravity loads only.

I will not proceed further with the girder design at this point, for the effects of lateral loading must also be considered. The moments determined here for the gravity loading will be combined with those from the lateral loading in the discussion in Section 13.3.

13.2 COLUMN DESIGN FOR GRAVITY LOADS

The four general cases for the columns are (Figure 13.8):

The interior column carrying primarily only axial gravity loads.

The intermediate exterior columns on the north and south sides carrying the ends of the interior girders and functioning as members of the perimeter bents for lateral resistance.

The intermediate exterior columns on the east and west sides carrying the ends of the column-line beams and functioning as members of the perimeter bents.

The corner columns carrying the ends of the spandrel beams and functioning as the end members in both peripheral bents.

Summations of the design loads for the columns may be done as illustrated for the steel column in Section 12.3. As all columns will be subjected to combinations of axial compression and bending, these gravity loads represent only the axial compression action. For the interior columns, the bending moments will be relatively low in comparison to the compression loads, and it is reasonable for a preliminary design to ignore bending effects and design for axial compression only. The usual minimum required eccentricity will likely provide for sufficient bending. On this basis, a trial design for one of the interior columns is shown in Figure 13.9. Design loads were obtained in a manner similar to that shown for the steel column in Section 12.3. A single size of 24 in. square is used for all three stories, a common practice permitting the reuse of column forming for cost savings. The service load capacities indicated may be compared with values obtained from the graphs in Figure B.6. Economy is also generally obtained with the use of low percentages of reinforcement when bending-moments are not a critical concern; the percentages shown in Figure 13.9 are minimal, but smaller column sizes could be used if floor space and planning problems are of major concern.

The interior column occurs at the location of the stairs and rest rooms, and it is possible that some form alteration may be made to allow the columns to fit more smoothly into the wall planning. This will add cost to the column construction, but is relatively easily achieved.

FIGURE 13.8 Framing situations for the columns and the column-line beams.

For the intermediate exterior column, there are four actions to consider:

1. The vertical compression induced by gravity.
2. Bending moment induced by the interior framing that intersects the wall column; the columns are what provides the end moments shown in Figures 13.5 and 13.7.
3. Bending moments in the plane of the wall induced by unbalanced conditions in the spandrel beams and girders.
4. Bending moments induced by the actions of the perimeter bents in resisting lateral loads.

For the corner column the situation is similar to that for the intermediate exterior column, that is, bending on both axes. The forms of the exterior columns as shown on the plan in Figure 13.1 have been established in anticipation of the major effects described. Further discussion of these columns will be deferred, however, until after an investigation of the situations of lateral loading.

13.3 DESIGN OF THE PERIMETER BENTS

The lateral force resisting systems for the concrete structure are shown in Figure 13.10*a*. For force in the east–west direction the resistive system con-

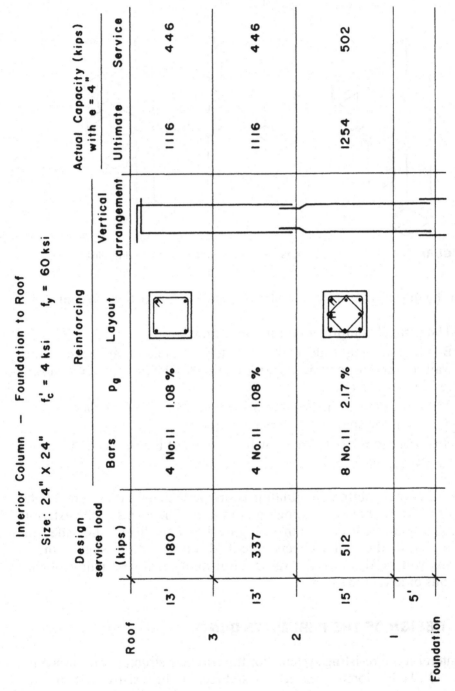

FIGURE 13.9 Summary of design for the interior column; gravity load only.

sists of the horizontal roof and floor slabs and the exterior bents (columns and spandrel beams) on the north and south sides. For force in the north-south direction the system utilizes the bents on the east and west sides. Actually other elements of the structural frame will also resist lateral force, but by widening the columns and deepening the spandrel beams in these bents, an increased stiffness will be produced; the stiffer bents will then tend to offer the most resistance to lateral movements. Design these stiffened bents for all the lateral force, ignoring the minor resistances offered by the other column and beam bents.

With the same building profile, the wind loads on this structure will be the same as those determined for the steel structure in Section 12.4. As in the example in that section, I will illustrate the design of the bents in only one direction, in this case the bents on the east and west sides. From the data given in Figure 12.9, the horizontal forces for the concrete bents are determined as follows:

$$H_1 = (195)(122)/2 = 11{,}895 \text{ lb}, \quad \text{say } 11.9 \text{ kips/bent}$$

$$H_2 = (234)(122)/2 = 14{,}274 \text{ lb}, \quad \text{say } 14.3 \text{ kips/bent}$$

$$H_3 = (227)(122)/2 = 13{,}847 \text{ lb}, \quad \text{say } 13.9 \text{ kips/bent}$$

Figure 13.10b shows a profile of one of the north–south bents with the bent loads.

For an approximate analysis, consider the individual stories of the bent to behave as shown in Figure 13.10c, with the columns developing an inflection point at their midheight points. Because the columns all move the same distance, the shear load in a single column may be assumed to be equal to the cantilever deflecting load and the individual shears to be proportionate to the stiffnesses of the columns. If the columns are all of equal stiffness in this case, the total load would be simply divided by four. However, the end columns are slightly less restrained as there is a beam on only one side. I will assume the net stiffness of the end columns to be one-half that of the interior columns. Thus the shear force in the end columns will be one-sixth of the load and that in the interior columns one-third of the load. The column shears for each of the three stories is thus as shown in Figure 13.11.

The column shear forces produce moments in the columns. With the column inflection points assumed at midheight, the moment produced by a single shear force is simply the product of the force and half the column height. These moments must be resisted by the end moments in the rigidly attached beams, and the actions are as shown in Figure 13.12. These effects due to the lateral loads may now be combined with the previously determined effects of gravity loads for an approximate design of the columns and beams.

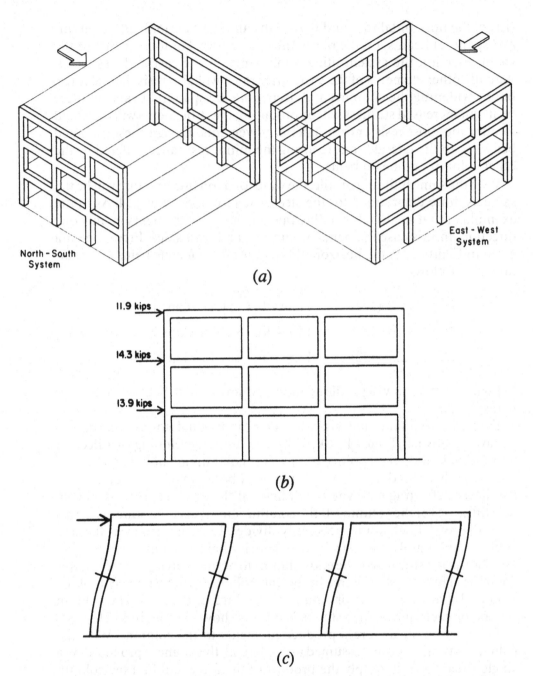

FIGURE 13.10 Actions of the moment-resisting bents for lateral loads: (*a*) Form of the perimeter bents. (*b*) Assumed lateral wind load for the north–south bent. (*c*) Assumed form of deformation of the bent stories.

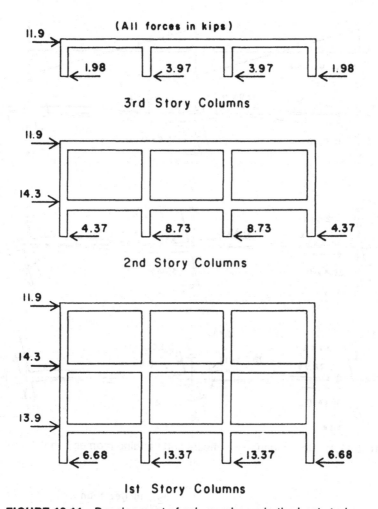

FIGURE 13.11 Development of column shears in the bent stories.

For the columns, combine the axial compression forces with any gravity-induced moments and first determine that the load condition without lateral effects is not critical. Then add the effects of the moments caused by lateral loading and investigate the combined loading condition, for which you may use the one-third increase in allowable stress. Gravity-induced beam moments are taken from Figure 13.7 and are assumed to induce column moments as shown in Figure 13.13 The summary of design conditions for the corner and interior column is shown in Table 13.1. The design values for axial load and moment and approximate sizes and reinforcing are shown in Figure 13.14. Column sizes and reinforcing were obtained from the tables in the *CRSI Handbook* (Ref. 6), using concrete with $f_c' = 4$ ksi and Grade 60 reinforcing.

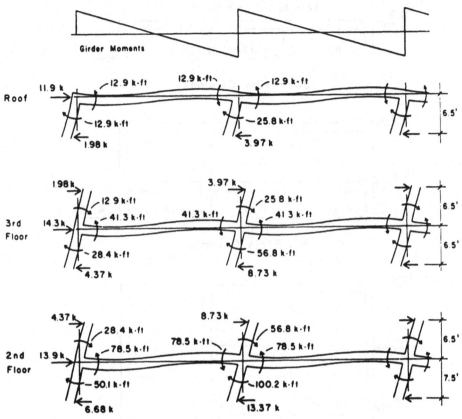

FIGURE 13.12 Investigation for shears and bending moments in the bent.

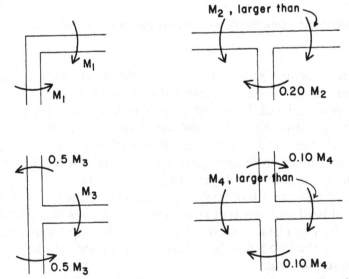

FIGURE 13.13 Assumptions for approximation of the distribution of bending moments in the bents due to gravity loading of the column-line beams and girders.

TABLE 13.1 Summary of Design Data for the Bent Columns

	Column	
	Intermediate	Corner
Axial gravity design load (kips)		
Third story	90.0	55.0
Second story	179.0	117.0
First story	277.0	176.0
Assumed gravity moment on bent axis (kip-ft) from Figures 13.13 and 13.7		
Third story	60.0	120.0
Second story	39.0	100.0
First story	39.0	100.0
Moment from lateral force (kip-ft) from Figure 13.12		
Third story	25.8	12.9
Second story	56.8	28.4
First story	100.2	50.1

	Intermediate Column					Corner Column				
	Axial Load	Moment	e	Column Dimensions	Reinforcement	Axial Load	Moment	e	Column Dimensions	Reinforcement
	(kips)	(kip-ft)	(in.)	(in.)	No. - Size	(kips)	(kip-ft)	(in.)	(in.)	No. - Size
Roof										
	90 X ¾ = 68	85.8 X ¾ = 64	11.3	20 X 28	6 - 9	55	120	35	20 X 24	6 - 10
3										
	179 X ¾ = 134	95.8 X ¾ = 72	6.5	20 X 28	6 - 9	117	100	13.6	20 X 24	6 - 10
2										
	277 X ¾ = 208	139.2 X ¾ = 105	6.1	20 X 28	6 - 10	176 X ¾ = 132	150.1 X ¾ = 113	10.3	20 X 24	6 - 10
1										

FIGURE 13.14 Design of the columns in the north–south bent for the combined loading.

The spandrel beams (or girders) must be designed for the combined shears and moments due to gravity and lateral effects. Using the values for gravity-induced moments from Figure 13.7 and the values for lateral load moments from Figure 13.12, the combined moment conditions are shown in Figure 13.15. For design we must consider both the gravity only moment and the combined effect. For the combined effect we use three-fourths of the total combined values to reflect the allowable stress increase of one-third.

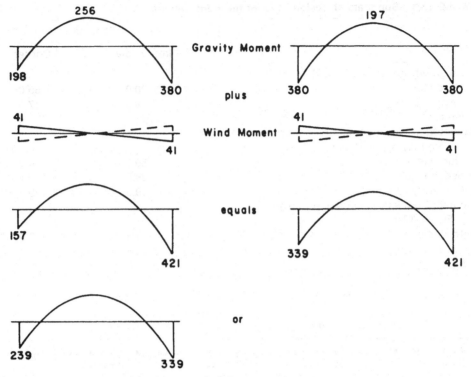

FIGURE 13.15 Summary of bending moments in the girders due to the combined loading.

Figure 13.16 presents a summary of the design of the reinforcing for the spandrel beam at the third floor. If the construction that was shown in Figure 13.3 is retained with the exposed spandrel beams, the beam is quite deep. Its width should be approximately the same as that of the column, without producing too massive a section. The section shown is probably adequate, but several additional considerations must be made.

For computation of the required steel areas, assume an effective depth of approximately 40 in. and use

$$A_s = \frac{M}{f_s jd} = \frac{M(12)}{24(0.9)(40)} = 0.0139M$$

Because the beam is so deep, it is advisable to use some longitudinal reinforcing at an intermediate height in the section, especially on the exposed face.

Shear design for the beams should also be done for the combined loading effects. The closed form for the shear reinforcing, as shown in Figure 13.16, is used for considerations of torsion as well as the necessity for tying the compressive reinforcing.

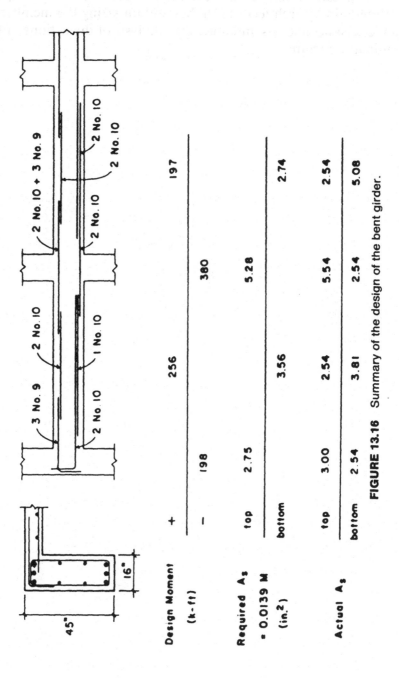

Design Moment (k-ft)		+	197	380	256	198
		−				
Required A_s = 0.0139 M (in.²)	top			5.28		2.75
	bottom		2.74		3.56	
Actual A_s	top		2.54	5.54	2.54	3.00
	bottom		5.08	2.54	3.81	2.54

FIGURE 13.16 Summary of the design of the bent girder.

With all of the approximations made, this should still be considered to be a very preliminary design for the beam. It should, however, be adequate for use in preliminary architectural studies and for sizing the members for a dynamic seismic analysis and a general analysis of the actions of the indeterminate structure.

PART V

BUILDING FIVE

14

GENERAL CONSIDERATIONS
FOR BUILDING FIVE

Structures are essentially three-dimensional. They may be formed with elements that are linear or planar, but the whole structure must exist in three dimensions. Buildings thus are spatial, and the structures that form them are necessarily spatial in form.

Planar trusses are common, although they are almost always elements in some ordered spatial arrangement. In many situations it is common to use bracing for planar trusses that itself constitutes a cross-bracing truss system. Building Five is a simple case of using trussing in a spatial system for the inherent three-dimensional stability it provides.

Trussing can be used in a more truly three-dimensional structural form—for example, the freestanding trussed tower. Although the slender tower takes a linear form overall, when fully developed with trussing, it becomes a true three-dimensional truss structure; it derives both its stability and structural strength from the three-dimensional arrangement of the truss members.

The more common example used to describe three-dimensional trussing is the two-way spanning truss structure, called a *space frame*.

14.1 THE BUILDING

This chapter presents some possibilities for developing the roof structure and the exterior walls for a medium-sized sports arena, one big enough for a swim stadium or a basketball court (see Figure 14.1).

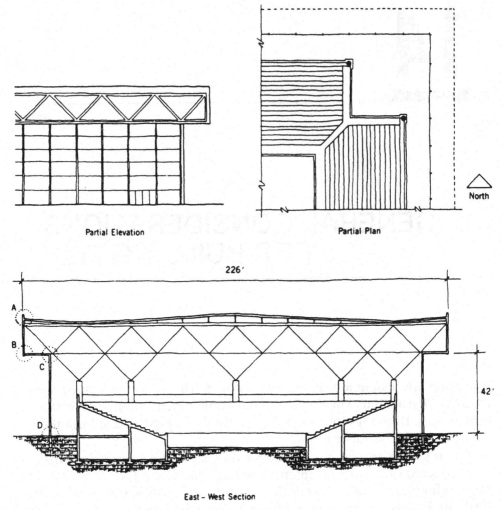

Partial Elevation   Partial Plan

△
North

226'

A
B
C
D

42'

East – West Section

FIGURE 14.1 Building Five: general form.

Options are strongly related to the desired building form. Functional planning requirements derive from the specific activities to be housed and from the seating, internal traffic, overhead clearance, and exit and entrance arrangements.

In spite of all the requirements, you can usually consider a range of alternatives for the general building plan and overall form. Your choices determine your flexibility in the building form. Choosing the truss sysem shown here, for example, commits you to a square plan and a flat roof profile, while selecting a dome limits you to a round plan.

In addition to the long-span structure in this case, you also must develop the 42-ft-high curtain wall. Braced laterally only at the top and bottom, this is a 42-ft span structure sustaining wind pressure as a loading. Even modest wind conditions produce pressures in the 20-psf range and thus require a major vertical mullion structure to span the 42 ft.

In this example the fascia of the roof trusses, the soffit of the overhang, and the curtain wall are all developed with products ordinarily available for curtain wall construction. The vertical span of the tall wall is developed by a series of custom-designed trusses.

As an exposed structure, the truss system is a major visual element of the building interior. The truss members define a pattern that is orderly and pervading. There will, however, most likely be many additional items overhead—within and possibly beneath the trusses—including

Leaders and other elements of the roof drainage system

Ducts and registers for the HVAC system

A general lighting system

Signs, scoreboards, and such

Elements of an audio system

Catwalks for access to the various equipment

To preserve some design order, these items should relate to the truss geometry and detailing, if possible. However, some independence is to be expected, especially with items installed or modified after the building is completed.

14.2 STRUCTURAL ALTERNATIVES

The general form of the construction for Building Five is shown in Fig. 14.2. Discussion for this building is limited to the development of the roof structure and the tall exterior glazed wall. Some proposed details for the exterior wall are shown in Figure 14.3; the details indicate the use of a standard, proprietary window wall system for the wall surface development, with a custom-designed steel support structure occurring mostly behind the wall and inside the building. The steel support structure for the wall is thus weather-protected by the glazed wall, which is ordinarily developed as a weather-resistive construction system.

The 168 ft clear span used here is definitely in the class of long-span structures, but it will not severely limit the options. Although a flat-spanning beam system is out, a one-way or two-way truss system is feasible for a flat span. Most other structural options generally involve some form other than a flat profile: domes, arches, shells, folded plates, suspended cables, cable-stayed systems, and pneumatic systems.

The structure shown in Figure 14.1 uses a two-way spanning steel truss system whose form is described as an *offset grid.* The basic planning of this

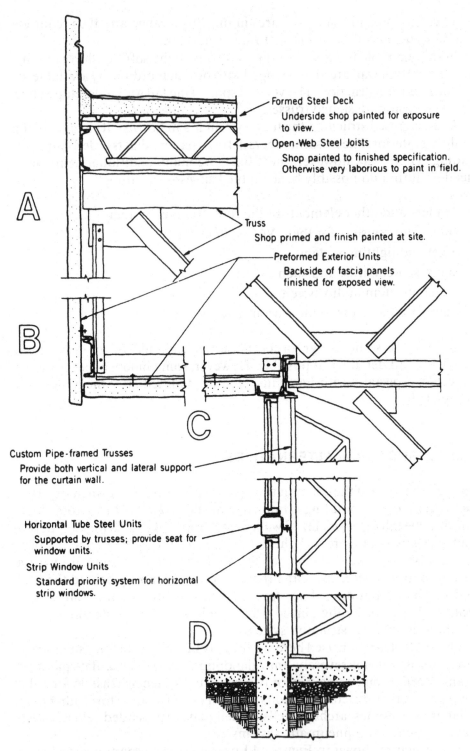

Formed Steel Deck

Underside shop painted for exposure to view.

Open-Web Steel Joists

Shop painted to finished specification. Otherwise very laborious to paint in field.

Truss

Shop primed and finish painted at site.

Preformed Exterior Units

Backside of fascia panels finished for exposed view.

Custom Pipe-framed Trusses

Provide both vertical and lateral support for the curtain wall.

Horizontal Tube Steel Units

Supported by trusses; provide seat for window units.

Strip Window Units

Standard priority system for horizontal strip windows.

FIGURE 14.2 Building Five: construction details.

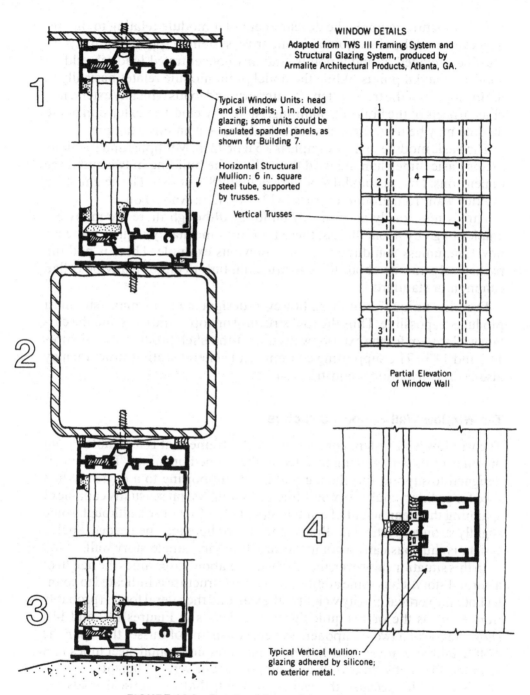

WINDOW DETAILS
Adapted from TWS III Framing System and
Structural Glazing System, produced by
Armalite Architectural Products, Atlanta, GA.

Typical Window Units: head
and sill details; 1 in. double
glazing; some units could be
insulated spandrel panels, as
shown for Building 7.

Horizontal Structural
Mullion: 6 in. square
steel tube, supported
by trusses.

Vertical Trusses

Partial Elevation
of Window Wall

Typical Vertical Mullion:
glazing adhered by silicone;
no exterior metal.

FIGURE 14.3 Details of the curtain wall system.

189

type of structure requires the development of a module relating to the frequency of nodal points (joints) in the truss system. Supports for the truss must be provided at nodal points, and any concentrated loads should be applied at nodal points. While the nodal point module relates basically to the formation of the truss system, its supports and loads typically extend it to other aspects of the building planning. At the extreme, this extension of the module may be used throughout the building—as in this example.

The basic module in the example is 3.5 ft or 42 in. Multiples and fractions of this basic dimension are used throughout the building in two and three dimensions. The truss nodal module is actually 8 × or 28 ft. The height of the exterior wall from ground to truss is 12 × or 42 ft. And so on.

This is not exactly an ordinary building, although there are many examples to go by. Nonetheless, these buildings inevitably require some innovation, unless you duplicate some previous example. There is nothing particularly unique about the construction, but it cannot really be called common or standard.

Even in "unique" buildings, however, designers use as many standard products as possible. Thus the roof structure on top of the truss and the curtain wall system for the exterior walls use off-the-shelf products (see Figures 14.2 and 14.3). The supporting columns and general seating structure may also be of conventional construction.

The Window Wall Support Structure

The window wall system consists of a metal frame that is assembled from modular units with glazing inserted in the framed openings. The frame is designed to support the glazing and is self-supporting to a degree. When used for ordinary multistory buildings, this wall system is usually capable of spanning the clear height of a single story (12 to 15 ft or so), although some slightly stronger vertical mullions than shown here may be needed. Such a system is thus basically a complete structure for a single-story wall.

In this situation, however, the 42 ft height is about three times the height of a normal story. Thus some additional support structure is indicated to counter both the vertical gravity weight of the wall and the lateral forces (probably from wind, as the wall is quite light). The details in Figures 14.2 and 14.3 show the use of a two-component system consisting of trusses that span the 42 ft height and are spaced 14 ft on center and horizontal steel tubes that span the 14 ft between trusses and support 7 ft of vertical height.

It is possible, because the wall is so light, that the vertical loads are actually carried by the closely spaced vertical mullions. In this case, the primary function of the horizontal steel tubes is to span between trusses and resist the wind forces on 14 ft of the wall. The tube shown here is certainly adequate for this task, although stiffness is probably more critical than bending stress: it is not really good for the wall construction to flip-flop during either seismic movements or fast changes in wind direction.

Although not shown in the drawings, the truss would probably be a so-called *delta truss*, where two chords oppose a single one, creating a triangular cross section (that is, it looks like the Greek capital letter delta). The delta truss has relatively high lateral stability, so you can use it without the usual required cross-bracing.

An example of such a window wall is shown in Figure 14.4.

Selection of the Truss System

This truss system can be produced from various available proprietary systems. When beginning to design such a structure, be sure to check whether you can obtain the necessary materials. A completely custom-designed structure of this kind requires enormous design time for basic development and planning of the structural form, development of nodal joint construction, and reliable investigation of structural behavior of the highly indeter-

FIGURE 14.4 Large trussed mullions used for a tall glazed wall similar to that for Building Five.

FIGURE 14.4   (*Continued*)

minate structure. Even developing assemblage and erection processes pose major design problems.

If the project deserves this effort, if the time for the design work is available, and if the budget can absorb the cost, the end result might justify the expenditures. But if what is really desired and needed can be obtained with available products and systems, you can save a lot of time and money.

Don't forget you need to identify the particular form and general nature of the truss structure. The square plan and general biaxial plan symmetry seem to indicate that you need a two-way spanning system (see Figure 14.1). The particular system form in this example is called an *offset grid* because the squares of the top chords are offset from those of the bottom chords so that the top chord nodules (joints) lie over the center of the bottom squares. As a result, there are no vertical web members and generally no vertical planar sets in the system.

14.3   GENERAL CONSIDERATIONS FOR TWO-WAY TRUSSES

An ordinary truss system consisting of a set of parallel, planar trusses may be turned into a two-way spanning system by connecting the parallel trusses with cross-trussing. However, this system lacks stability in the horizontal plane. If viewed in plan, the two sets of vertical, planar trusses form rectangles, which do not have the inherent triangulation needed for stability in the horizontal plane. Of course, you can rectify this situation by adding horizontal trussing.

Whereas the planar triangle is the basic unit of planar trusses, the tetrahedron is the basic unit of the spatial truss (see Figure 14.5). If a three-dimensional system is developed with three orthoginal planes (x,y,z coordinate system), the system basically defines rectangles in each plane and cubical forms in three dimensions. Triangulation of each orthogonal plane actually produces sets of tetrehedrae in space, as shown in Figure 14.5.

When you use the mutually perpendicular vertical, planar truss system, you can easily form squared corners at the edge of the system and in relation to plan layouts beneath it. For example, in the roof structure of the dining hall at the Air Force Academy in Colorado Springs, Colorado, vertical trusses are used to form a square plan with a clear interior span of 266 ft. The roof edge is cleanly formed by the basic truss system, and the exterior walls meet the bottom of the truss at natural locations of the truss chords. Although you can't see it in the building's exterior, the truss system is visible in the modules used for the roof fascia and the exterior walls (see Figures 14.6 and 14.7).

A purer form for the spatial truss derives from the basic spatial triangulation of the tetrahedron. If all the edges of the tetrahedron (or truss members) are equal in length, the solid form described is not orthogonal; that is, it does not describe the usual x,y,z system of mutually perpendicular planes (see Figure 14.8). If used for a two-way spanning truss system, a flat structure in the horizontal plane may be developed, but it will not inherently develop rectangles in plan: its natural plan form will be in multiples of triangles, diamonds, and hexagons.

For small trusses, the fully triangulated system can save you money because it can be fully formed with all members of the same length and a single, simple joint can be used for most of the truss system.

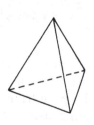
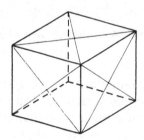

single tetrahedron   the trussed box

FIGURE 14.5 Three-dimensional trussing: the single tetrahedron and the developed cube.

FIGURE 14.6 Large roof structure with intersecting vertical-planar steel trusses supported on four sides by columns. Air Force Academy dining facility, Colorado Springs, Colorado. Architects and engineers: Skidmore, Owings, and Merrill, Chicago.

FIGURE 14.7 Exterior view of the Air Force Academy dining facility.

The pure triangulated space system, however, allows you to form truly three-dimensional structures and not just flat, two-way spanning systems.

A compromise form—somewhere between the pure triangular and the orthogonal systems—is the offset grid. This system consists of two horizontal planes, each constituted as square, rectangular grids. However, the upper plane grid (top chord plan) has its grid intersections located over the centers of the grids in the lower plane (bottom chord plan). The truss web members are arranged to connect the top grid intersections with the bottom grid intersections. The typical joint therefore consists of the meeting of four chord members and four web diagonal members. The web members describe vertical planes in diagonal plan directions but do not form vertical planes with the chords.

The offset grid permits development of square plan layouts (usually preferred by designers), while retaining some features of the triangulated system (members the same length, joints the same form). Some details of offset grid systems are shown elsewhere in this chapter. An example system is illustrated in Section 15.2.

Truss geometry must relate reasonably to the purpose of the structure—often the forming of a building's roof. For less constricting design situations, such as the forming of a canopy, a structural sculpture, or a theme structure, the space frame may be developed more dramatically.

You can use spatial trussing to produce just about any form of structural unit, including columns, trussed mullions, freestanding towers, beams, sur-

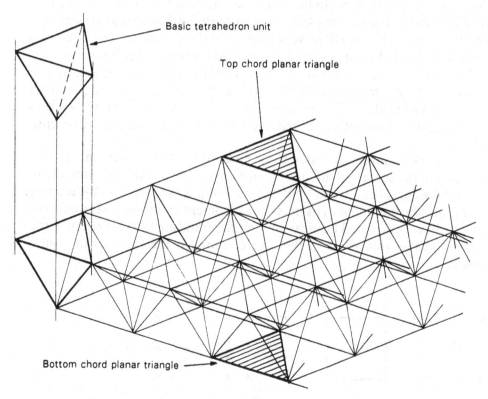

FIGURE 14.8 Two-way truss system developed with the tetrahedron as a basic geometric unit.

faces of large multiplaned structures, and even arched surfaces formed as vaults (cylinders) or domes.

In many situations trussing offers a practical, economically efficient structural solution. It also offers extreme lightness and a visual openness, which are often major design considerations. Light transmission, for example, may be a major factor, not just aesthetically but functionally.

Span and Support Considerations

When planning structures that have spatial trussing, you must pay attention to the nature of the supports if you want to realize optimal use of the two-way action. The locations of supports define not only the size of spans to be achieved, but many aspects of the structural behavior of the two-way truss system.

Figure 14.9 shows possible support systems for a single square panel of a two-way spanning system. In Figure 14.9*a* support is provided by four columns placed at the outside corners, resulting in a maximum span condition for the system's interior portion. This form of support also results in a very high shear condition in the single quadrant of the corner and requires that the edges act as one-way spanning supports for the two-way system. As a result, the edge chords will be very heavy and the corner web members will be heavily loaded for transfer of the vertical force to the columns.

Figures 14.9*b* and *c* show supports that eliminate the edge spanning and corner shear by providing either bearing walls or closely spaced perimeter columns. The trade-off is a lower truss cost for higher costs for the support system and its foundations. Such support is more restrictive architecturally.

Figure 14.9*d* shows an interesting possibility. The placement of four columns in the centers of the sides requires a considerable edge structure to achieve the corner cantilevers, but it actually reduces the span of the interior system and further reduces its maximum moment by the overhang effect of the cantilever corners. The clear span of the building interior remains the same, and the high shear at the four columns is the same as in Figure 14.9*a*.

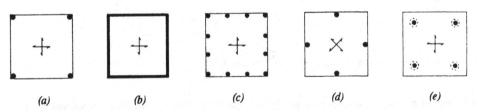

(a)   *(b)*   *(c)*   *(d)*   *(e)*

FIGURE 14.9   Support variations for a single, two-way spanning structure.

For an ideal column shear condition, the solution shown in Figure 14.9 places the columns inside the edges. This provides a wider edge spanning strip as well as a total perimeter of the truss system at each column with shear divided between more truss members. It reduces the clear interior span to something less than the full width of the truss. If the exterior wall is at the roof edge, the interior columns may be intrusive in the plan. However, if a roof overhang is desired, the wall may be placed at the column lines and the structure will relate well to the architectural plan. This is the solution illustrated in Figure 14.7 and in the example designs for Building Five.

Single units of a two-way spanning system should describe a square as closely as possible. Even though the assemblage has a potential for two-way spanning, if the support arrangement provides oblong units, the structural action may be essentially one-way in function. Figure 14.10 shows three forms for a span unit, with sides in ratios of 1:1, 1:1.5, and 1:2. If the system is otherwise fully symmetrical, the square unit will share the spanning effort equally in each direction.

If the ratio of sides gets as high as 1:1.5, the shorter span becomes much stiffer for deflection and will attract as much as 75% of the total load. This ratio is usually the maximum one for any practical consideration of two-way action.

If the ratio of the sides gets as high as 1:2, scarcely any load will be carried in the long direction, except for that near the ends of the plan unit, adjacent to the short sides. Two-way action relates generally to the development of two-way curvature (domed or dishlike form), and it should be clear that the long, narrow unit will bend mostly in single, archlike form.

Two-way systems—especially those with multiple spans—are often supported on columns. This generally results in a high shear condition at the supporting columns and various means are used to relieve the interior force concentrations in the spanning structure. A common solution for the two-way spanning concrete flat slab has the effective perimeter size of the

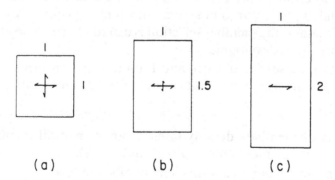

(a)   (b)   (c)

FIGURE 14.10 Variations of lengths of sides for a single, two-way spanning system.

column extended by an enlarged top (called a *column capital*) and the strength of the slab increased locally by a thickened portion at the column (called a *drop panel*). These elements have their analogous counterparts in systems using two-way trussing.

Figure 14.11 shows the structure for a large offset grid truss, supported only by four columns. On the top of each column is a cross-shaped member that extends to pick up four truss joints; ordinarily the single column would be able to support only a single joint. In this case, the shear in the truss is shared by four times as many web members as it would be without the column-top device. Similar application is shown in Figure 14.12.

In some truss systems it is possible to develop something analogous to the drop panel in the concrete slab. Thus an additional layer of trussing, or a single dropped-down unit, may be provided to achieve the local strengthening within the truss system itself.

Joints and System Assembly

Many of the issues to consider in developing joints for spatial truss systems are essentially the same as those considered for planar truss systems: selection of materials, shape of truss member, arrangement of members at a joint, and magnitude of loads. These primary decisions preface selection of jointing methods (welding, bolting, nailing, and so on) and the use of intermediate devices (gusset plates, nodal units, and so on).

For the spatial system two additional concerns are significant. The first is that most joints are of a three-dimensional character, relating to the specific geometry of the system. This generally calls for a somewhat more complex joint development than the simple alternatives often possible for planar trusses. A second concern is that the typically large number of joints requires a relatively simple, economically feasible joint construction.

Proprietary truss systems are often characterized by a special, clever jointing system that accommodates the variety of joints for the system, is inexpensive when mass-produced, and can be quickly and easily assembled. Although a particular joint system may relate specifically to a single truss member shape, the member selection is usually not as critical a design problem as the joint development.

Most spatial trusses are of steel, and I am mostly concerned here with methods for assemblage of steel frameworks. The primary jointing methods are:

By welding of members directly to each other or of all members to a primary joint element (gusset, node, and so on).

By bolting, most likely to a joint element of some form

By direct connection with threaded, snap-in, or other attachments

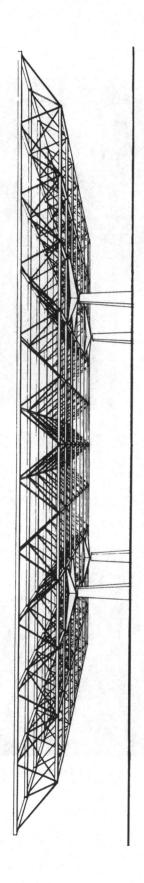

FIGURE 14.11 Large two-way spanning truss system supported on only four columns. Columns have a four-fingered, cantilevered top unit that picks up four truss nodes at the bottom chord. High school gym, Pekin, Illinois. Architects: Foley, Hackler, Thompson and Lee, Peoria, Illinois. Engineers: The Engineers Collaborative, Chicago, Illinois.

FIGURE 14.12 Offset-grid truss used for a moderate-span length, with extended support units on the column tops.

As with other truss systems, it is often possible and desirable to prepare large units of the total structure in a fabricating shop and transport them to the building site where they are bolted together and to their supports. Careful design of the field joints and of the shapes and sizes of shop-fabricated units can minimize erection problems. This procedure is more often the case with relatively large-scale trusses of individual design than with proprietary systems, which generally consist of individual members whose joints are individually field-assembled.

Joints may also be developed to facilitate other functions, besides that of the truss assemblage. Figure 14.13 shows a jointing system used for an offset

FIGURE 14.13 Offset-grid truss used for a roof structure for a shoppinig center concourse area. Truss system provides support for a wood roof system and a series of skylights.

grid system. Joints are achieved by bolting of steel plate elements, some occurring as attachments to the member ends and others as bent plate node units. An additional function required here is the support of an ordinary timber frame roof and plank deck. A unit on top of the top chord joints supports the wood roof system. A further consideration in this structure is the facilitation of slope for the roof surface. This is done by varying the height of the top chord support element, permitting the truss system to be built dead flat in the horizontal plane: a simple pragmatic solution for an otherwise quite sophisticated system.

Joints may also need to accommodate thermal expansion, seismic separation, or some specific controlled structural action (for example, pinned joint response to avoid transfer of bending moments). Support of suspended elements is often a requirement as well. Elements supported may relate to the module of the truss as defined by the joints or the member spacing.

A special joint is one that occurs at a support for the truss system. Beyond the usual requirement for direct compression bearing, there may be need for tension uplift resistance or lateral force for wind or seismic actions. It may also be necessary to achieve some real pin joint response to avoid transfer of bending to column tops as the truss deflects. Varying chord length due to thermal expansion or live load stress changes may also present some problems. Of course, the designer must consider the composition of the supporting structure as well.

15

DESIGN OF THE
ROOF STRUCTURE

The general form of the building as shown in Figure 14.1 can be developed with a considerable variety of truss systems for the roof. In this chapter I discuss several options for the truss system, all of which can achieve the form proposed for this building.

15.1 DEVELOPMENT OF THE ROOF INFILL SYSTEM

All of the schemes for the roof truss system shown here involve providing a grid of truss chords at 28 ft centers in both directions at the general level of the roof surface (top of the trusses). This provides a roof support system but does not develop a roof as such. Something else must be done to achieve the infill surface-developing construction, with its top surface consisting of some form of water-resistive roofing.

While design could feature an imaginative two-way spanning system for this structure, various simple systems will suffice. Figure 15.1 shows a very simple system consisting of simple-span steel open-web joists and formed sheet steel decking. For the simplest construction, the joists would use only the truss chords in one direction for support, creating a somewhat unsymmetrical loading for the two-way system. Except for the related connection details, however, this would most likely not affect the basic development of the larger truss system. The steel deck could be exposed on its underside,

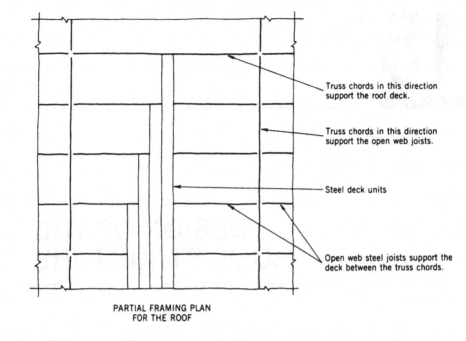

PARTIAL FRAMING PLAN
FOR THE ROOF

Truss chords in this direction support the roof deck.

Truss chords in this direction support the open web joists.

Steel deck units

Open web steel joists support the deck between the truss chords.

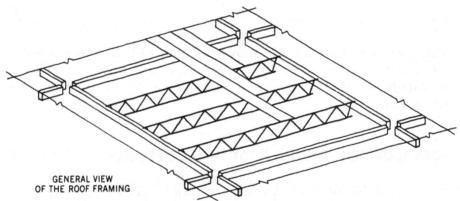

GENERAL VIEW
OF THE ROOF FRAMING

FIGURE 15.1   The infill roof framing system.

along with the open-web joists. However, for extra thermal insulation and possibly for sound control, it may be desirable to provide a ceiling surface at the bottom of the open-web joists.

Roof drainage is also a problem that relates to the form of the two-way truss system. If the open-web joists are placed as shown in Figure 15.1 (the usual installation), roof drainage might necessitate sloping the two-way truss system top chords. This would suggest using additional elements between the two-way trusses and the roof infill structure to permit the two-way system to be flat on top—surely a simplifying condition for the truss system.

This basic infill system is assumed to be reasonable for all of the following proposed alternatives for the two-way system, except for the one described in Section 15.5, which uses a different system for the center portion of the roof.

15.2   THE TWO-WAY OFFSET GRID SYSTEM

Figures 15.2 and 15.3 show the form of an offset grid system for the truss roof for Building Five. The plan shows the placement of the grid system resulting in the locations of supports beneath top chord joints. This is the general form indicated in the drawings in Figures 14.1 through 14.3.

The drawings in Figure 14.1 indicate the use of three columns on each side of the structure, providing a total of 12 supports for the truss system. The tops of the columns are dropped below the spanning truss to permit the use of a pyramidal module of four struts between the top of the column and the bottom chords of the truss system. This reduces the maximum shear required by the truss interior members, with the entire gravity load being shared by 12 ×

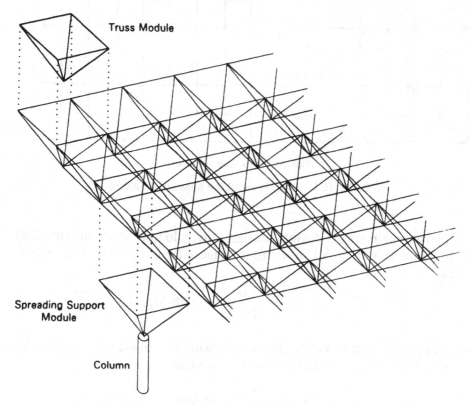

Truss Module

Spreading Support
Module

Column

FIGURE 15.2   General form of the offset grid system.

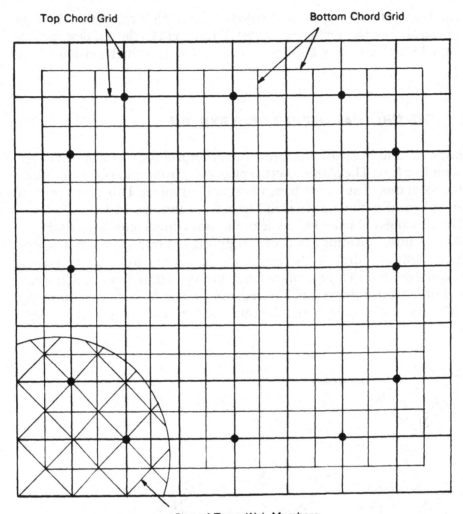

Top Chord Grid

Bottom Chord Grid

Plan of Truss Web Members

FIGURE 15.3 Plan layout of the offset grid system.

4 = 48 truss members. If the total design gravity load is approximately 100 psf (0.1 kips/ft²), the load in a single diagonal column strut is approximately

$$C = \frac{(226)^2(0.1)}{48} = 106 \text{ kips}$$

and since each strut picks up a truss node with four interior diagonals, the maximum internal force in the interior diagonals is

$$C = \frac{106}{4} = 26 \text{ kips}$$

If steel pipe is selected for the 28-ft-long members, possible choices from tables in the *AISC Manual* (Ref. 2) would be

6 in. standard for the truss member (37 kip capacity)
8 in. extra strong for the strut (137 kip capacity)

The relatively closely spaced edge columns plus the struts constitute an almost continuous edge support for the truss system, with only a minor edge cantilever. Thus the spanning task is essentially that of a simple beam span in two directions. For an approximation, consider the span in each direction to carry half the load. Taking half the clear span width of 168 ft as a middle strip, the total load for design of the "simple beam" is

$$W = (\text{span width})(\text{span length}) (0.1 \text{ kips/ft}^2)$$

$$= (84)(168)(0.1) = 1411 \text{ kips}$$

and the simple beam moment at midspan is

$$M = \frac{WL}{8} = \frac{(1411)(168)}{8} = 29631 \text{ k-ft}$$

Sharing this with three top chord members in the middle strip, and assuming a center-to-center chord depth of approximately 19 ft, the force in a single chord is

$$C = \frac{29631}{(3)(19)} = 520 \text{ kips}$$

If the compression chord is unbraced for its 28 ft length, this is beyond the capacity of a pipe (at least from the AISC Tables) but could be achieved with a W-shape (W 12 × 136 with F_y of 50 ksi) or a pair of thick angles (8 × 8 × 1 in. with F_y = 50 ksi).

The critical problem for the bottom chords will be development of joints at nodes or splices. It is unlikely that chord members would be more than a single module (28 ft) long, so each joint must be fully developed with welds or bolts.

These are not record sizes for large steel structures, but it suggests some strong efforts be made to reduce the design loads (or going for the lightest possible general roof construction, for example). It may also suggest a variety of sizes be used in the truss, with some minimal members used for low-stress situations.

This is a highly indeterminate structure, although its symmetry and the availability of computer-aided procedures for investigation make its design accessible to most professional structural designers.

15.3 THE TWO-WAY VERTICAL PLANAR TRUSS SYSTEM

A second possibility for the truss form is shown in Figure 15.4, consisting of perpendicular, intersecting sets of vertical planar trusses. In this system the top chord grid squares are directly above the bottom chord grid squares.

Figure 15.5 shows a plan layout for this system in which the vertical truss planes are offset from the columns. The principal structural reason for this is to permit the use of the spread unit at the column, similar to the one used in the example in Section 15.2. This unit does not relate to the basic truss system form as was the case in that example, but it could take various shapes to fulfill its task.

Approximation of the chord forces for this two-way spanning system could also be made in a manner similar to that for the offset grid structure. An advantage here is the possibility to use interior vertical members to reduce the lateral support problem for the chords. With the addition of some

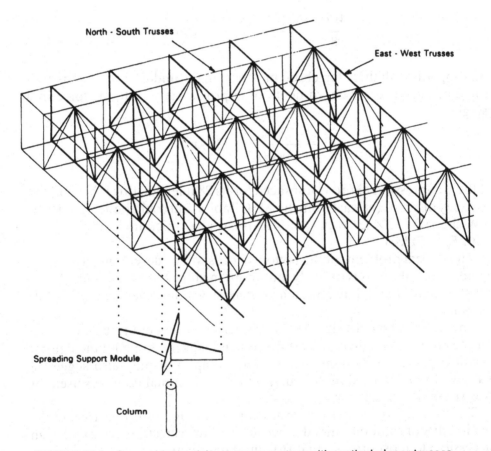

FIGURE 15.4 General form of the two-way system with vertical planar trusses.

Edge Carrying Trusses   Two-Way Center Truss Grid System

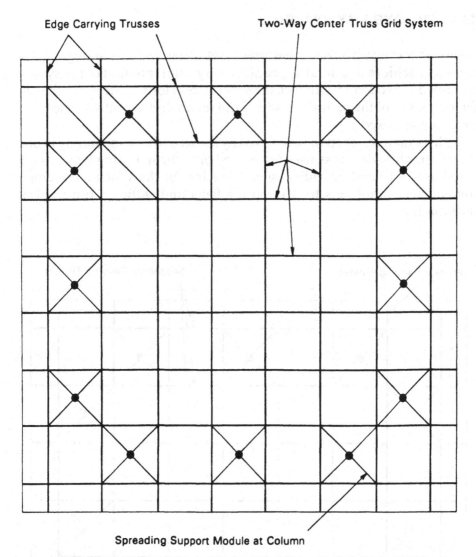

Spreading Support Module at Column

FIGURE 15.5   Plan layout of the system with vertical planar trusses.

members, this would have been possible in the offset grid system too, but not achieved so neatly. Using the same force approximations as determined for the chords in the offset example, but with unsupported lengths of only 14 ft, some considerably smaller members will be obtained.

For both of these two-way systems, planning for the erection includes determining what size and shape unit can be assembled in the shop and transported to the site and what temporary support must be provided. These decisions may be influenced by the design of the truss jointing details.

15.4 THE ONE-WAY SYSTEM

Figure 15.6 shows a system that uses a set of one-way spanning, planar trusses to achieve a general appearance very similar to that of the system described in Section 15.3. In fact, the general truss form is identical; the difference lies in of the manner in which trusses are individually formed and joints are achieved.

In this example, the span is achieved by the set of trusses spanning in one direction, while the cross-trussing is used only for spanning between the carrying trusses and providing lateral bracing for the system. The cross-trussing also cantilevers to develop the facia and soffit on two sides of the building.

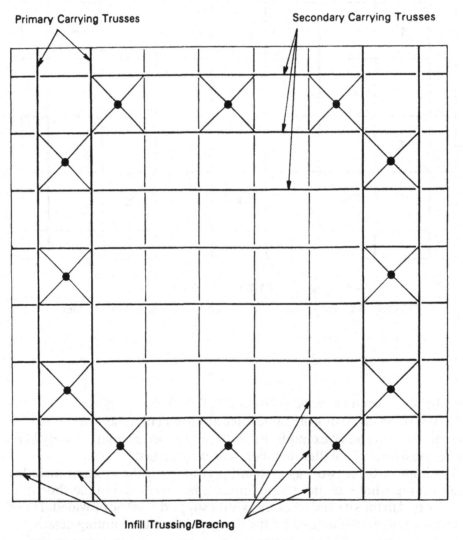

FIGURE 15.6 Plan layout of the system with intersecting one-way trusses.

Because the main trusses are simple planar, determinate trusses, this system lends itself to relatively simple design procedures. Perhaps this is not a compelling reason for choosing this scheme, but it is food for thought when you consider the complexity of investigations of highly indeterminate systems.

The 28 ft on center carrying trusses will be slightly heavier than the trusses in the preceding scheme since they take more of the load than in two-way structures. This is compensated for by the minor structural demands for the cross-trussing system. The designer would have to make a philosophical decision about how far to go to make the structure appear to be symmetrical in the otherwise biaxially symmetrical building. In reality, however, if no

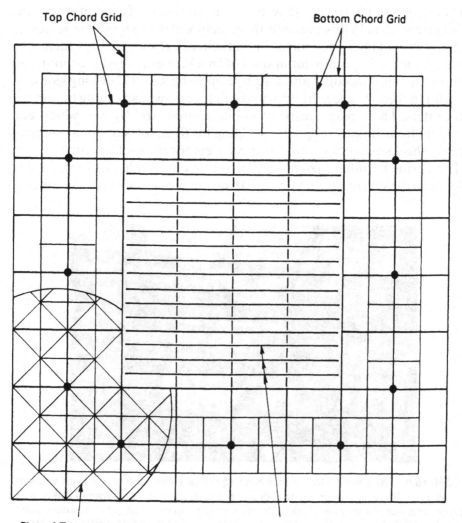

FIGURE 15.7 Plan layout for the composite system with a perimeter offset grid and a center one-way infill.

effort is made, most nonprofessionals will probably never notice the lack of symmetry. This is most likely true of most of the subtlety designers put into buildings; it is lost on all but fellow professionals.

A principal potential advantage for this scheme is its simplified assemblage and erection. Single carrying trusses may be erected in one piece with very little temporary support necessary. Once two that straddle a column are in place, the development of the cross-trussing can begin, serving both temporary and permanent bracing functions. This may be the single critical factor favoring this scheme.

15.5 THE COMPOSITE SYSTEM

The drawings in Figure 15.7 show a system consisting of a variation on the offset grid system. In this example, the structure at the four sides is achieved with a system of the same form as in that scheme. However, this structure is limited to achieving a maximum span of 56 ft between the edge columns. It defines the building edges and a square opening with 112-ft-long sides.

To fill the center space, the drawings call for one-way spanning trusses on 7 ft centers. These may consist of manufactured steel trusses, which are available from various suppliers. Bridging for these trusses plus a one-way formed sheet steel deck completes the system for the roof structure.

The perimeter offset system here could be almost identical to that developed in the scheme in Section 15.2. However the lack of biaxial symmetry

(a)

FIGURE 15.8 Truss system for the roof of a large square enclosed space. Uses one-way trusses and infill cross-trussing. The roof surface is developed with pyramidal skylights. (a) A general view of the system looking toward the north glazed wall. (b) The main trusses, formed with W-shape members. (c) A view of the roof skylights. Hyatt Regency Hotel, Chicago, Illinois.

(b)

(c)

FIGURE 15.8 *(Continued)*

causes the trusses and supports on two sides to be slightly more heavily loaded.

Figure 15.8 shows some views of a one-way spanning system used for the roof of a large open space at the lobby level of a hotel. The trusses support a glazed system consisting of pyramidal skylights. An additional consideration was the desire for an open view on the street side, which faces north and has among other things, a view of the John Hancock building on the horizon. Although the space is essentially square, the one-way spanning system was chosen for various reasons, including the ability to achieve the very light mullion system for the north-facing wall. The main spanning trusses are quite heavy, with large W shapes for members, but the system is actually quite light-appearing due to the fully glazed roof above.

APPENDIX A

DESIGN LOADS

Structural tasks, which are defined primarily in terms of the loading conditions imposed on the structure, are derived by some combination of experience, common sense, and various regulations established by building codes and industry standards.

A.1 DEAD LOADS

Dead load is the building's weight; that is, dead loads are the weights of the materials that make up the walls, partitions, columns, framing, floors, roofs, ceilings, and so on. Dead loads, which are due to gravity, result in downward vertical forces. Thus, when designing a beam, you must allow for the weight of the beam itself. Table A.1, which lists the weights of many construction materials, may be used to compute dead loads.

Dead load is a permanent load, once the building construction is completed, unless remodeling or rearrangement of the construction occurs. Because of its permanence, dead load requires certain considerations by the designer, such as the following:

1. It is always included in design loading combinations, except for investigations of singular effects, such as deflections due to live load only.

TABLE A.1 Weights of Building Construction

	lb/ft^2	kN/m^2
Roofs		
3-ply ready roofing (roll, composition)	1	0.05
3-ply felt and gravel	5.5	0.26
5-ply felt and gravel	6.5	0.31
Shingles		
Wood	2	0.10
Asphalt	2–3	0.10–0.15
Clay tile	9–12	0.43–0.58
Concrete tile	8–12	0.38–0.58
Slate, $\frac{1}{4}$ in.	10	0.48
Fiber glass	2–3	0.10–0.15
Aluminum	1	0.05
Steel	2	0.10
Insulation		
Fiber glass batts	0.5	0.025
Rigid foam plastic	1.5	0.075
Foamed concrete, mineral aggregate	2.5/in.	0.0047/mm
Wood rafters		
2 × 6 at 24 in.	1.0	0.05
2 × 8 at 24 in.	1.4	0.07
2 × 10 at 24 in.	1.7	0.08
2 × 12 at 24 in.	2.1	0.10
Steel deck, painted		
22 ga	1.6	0.08
20 ga	2.0	0.10
18 ga	2.6	0.13
Skylight		
Glass with steel frame	6–10	0.29–0.48
Plastic with aluminum frame	3–6	0.15–0.29
Plywood or softwood board sheathing	3.0/in.	0.0057/mm
Ceilings		
Suspended steel channels	1	0.05
Lath		
Steel mesh	0.5	0.025
Gypsum board, $\frac{1}{2}$ in.	2	0.10
Fiber tile	1	0.05
Drywall, gypsum board, $\frac{1}{2}$ in.	2.5	0.12
Plaster		
Gypsum, acoustic	5	0.24
Cement	8.5	0.41
Suspended lighting and air distribution	3	0.15
Systems, average		
Floors		
Hardwood, $\frac{1}{2}$ in.	2.5	0.12
Vinyl tile, $\frac{1}{8}$ in.	1.5	0.07

TABLE A.1 (*Continued*)

	lb/ft^2	kN/m^2
Asphalt mastic	12/in.	0.023/mm
Ceramic tile		
$\frac{3}{4}$ in.	10	0.48
Thin set	5	0.24
Fiberboard underlay, $\frac{5}{8}$ in.	3	0.15
Carpet and pad, average	3	0.15
Timber deck	2.5/in.	0.0047/mm
Steel deck, stone concrete fill, average	35–40	1.68–1.92
Concrete deck, stone aggregate	12.5/in.	0.024/mm
Wood joists		
2 × 8 at 16 in.	2.1	0.10
2 × 10 at 16 in.	2.6	0.13
2 × 12 at 16 in.	3.2	0.16
Lightweight concrete fill	8.0/in.	0.015/mm
Walls		
2 × 4 studs at 16 in., average	2	0.10
Steel studs at 16 in., average	4	0.20
Lath, plaster; *see Ceilings*		
Gypsum drywall, $\frac{5}{8}$ in. single	2.5	0.12
Stucco, $\frac{7}{8}$ in., or wire and paper or felt	10	0.48
Windows, average, glazing + frame		
Small pane, single glazing, wood or metal frame	5	0.24
Large pane, single glazing, wood or metal frame	8	0.38
Increase for double glazing	2–3	0.10–0.15
Curtain walls, manufactured units	10–15	0.48–0.72
Brick veneer		
4 in., mortar joints	40	1.92
$\frac{1}{2}$ in., mastic	10	0.48
Concrete block		
Lightweight, unreinforced—4 in.	20	0.96
6 in.	25	1.20
8 in.	30	1.44
Heavy, reinforced, grouted—6 in.	45	2.15
8 in.	60	2.87
12 in.	85	4.07

2. Over time, it causes sag and requires reduction of design stresses in wood structures, produces creep effects in concrete structures, and so on.

3. It contributes some unique responses, such as stabilizing effects that resist uplift and overturn due to wind forces.

A.2   LIVE LOADS

Live loads technically include all the nonpermanent loadings that can occur, besides the dead loads. However, the term usually refers only to the vertical gravity loadings on roof and floor surfaces. These loads occur in combination with the dead loads but are generally random in character and must be dealt with as potential contributors to various loading combinations.

Roof Loads

Roofs are designed for not only the dead loads they support, but also a uniformly distributed live load that includes snow accumulation and the general loadings that occur during roof construction and maintenance. Snow loads are based on local snowfalls and are specified by local building codes.

Table A.2 gives the minimum roof live load requirements specified by the 1994 edition of the *Uniform Building Code.* Note the adjustments for roof slope and for the total area of roof surface supported by a structural element. The latter accounts for the increase in probability of the lack of total surface loading as the size of the surface area increases.

Roof surfaces must also be designed for wind pressure; the magnitude and manner of application are specified by local building codes. For very light roof construction, be careful that the upward (suction) effect of the wind does not exceed the dead load and result in a net upward lifting force.

Although the term *flat roof* is often used, all roofs must be designed for some water drainage. The minimum required pitch is usually $\frac{1}{4}$ in./ft, or a slope of approximately 1:50. With roof surfaces that are this close to flat, a potential problem is *ponding*, or when the weight of water on the surface causes deflection of the supporting structure, which in turn allows for more water accumulation (in a pond), causing more deflection, and so on; the result is an accelerated collapse condition.

Floor Loads

The live load on a floor represents the probable effects created by occupancy. It includes the weights of human occupants, furniture, equipment, stored materials, and so on. All building codes provide minimum live loads to be used when designing buildings for various occupancies. Since different codes specify different live loads, always use the local code. Table A.3 contains values for floor live loads as given by the 1994 edition of the *Uniform Building Code.*

Although expressed as uniform loads, code-required values are usually large enough to account for ordinary concentrations. For offices, parking

TABLE A.2 Minimum Roof Live Loads[1] (UBC Table 16-C)

ROOF SLOPE	METHOD 1			METHOD 2		
	Tributary Loaded Area in Square Feet for Any Structural Member			Uniform Load[2]	Rate of Reduction r (percentage)	Maximum Reduction R (percentage)
	0 to 200	201 to 600	Over 600			
	× 0.0929 for m² × 0.0479 for kN/m²					
1. Flat[3] or rise less than 4 units vertical in 12 units horizontal (33.3% slope). Arch or dome with rise less than one eighth of span	20	16	12	20	.08	40
2. Rise 4 units vertical to less than 12 units vertical in 12 units horizontal (33% to less than 100% slope). Arch or dome with rise one eighth of span to less than three eighths of span	16	14	12	16	.06	25
3. Rise 12 units vertical in 12 units horizontal (100% slope) and greater. Arch or dome with rise three eighths of span or greater	12	12	12	12		
4. Awnings except cloth covered[4]	5	5	5	5	No reductions permitted	
5. Greenhouses, lath houses and agricultural buildings[5]	10	10	10	10		

[1]Where snow loads occur, the roof structure shall be designed for such loads as determined by the building official. See Section 1605.4. For special-purpose roofs, see Section 1605.5.

[2]See Section 1606 for live load reductions. The rate of reduction r in Section 1606 Formula (6-1) shall be as indicated in the table. The maximum reduction R shall not exceed the value indicated in the table.

[3]A flat roof is any roof with a slope of less than $1/4$ unit vertical in 12 units horizontal (2% slope). The live load for flat roofs is in addition to the ponding load required by Section 1605.6.

[4]As defined in Section 3206.

[5]See Section 1605.5 for concentrated load requirements for greenhouse roof members.

Source: Uniform Building Code, 1994 ed. Reproduced with permission of the publisher, International Conference of Building Officials.

garages, and some other occupancies, codes often require the consideration of a specified concentrated load as well as the distributed loading. If buildings are to contain heavy machinery, stored materials, or other contents of unusual weight, these loads must be provided for individually in the design.

When structural framing members support large areas, most codes allow some reduction in the total live load used for design. In the case of roof loads, these reductions are incorporated into Table A.2. The following method is

TABLE A.3 Minimum Floor Live Loads (UBC Table 16-A)

USE OR OCCUPANCY		UNIFORM LOAD[1] (pounds)	CONCENTRATED LOAD (pounds)
Category	Description	× 0.004 48 for kN	
1. Access floor systems	Office use	50	2,000[2]
	Computer use	100	2,000[2]
2. Armories		150	0
3. Assembly areas[3] and auditoriums and balconies therewith	Fixed seating areas	50	0
	Movable seating and other areas	100	0
	Stage areas and enclosed platforms	125	0
4. Cornices and marquees		60[4]	0
5. Exit facilities[5]		100	0[6]
6. Garages	General storage and/or repair	100	[7]
	Private or pleasure-type motor vehicle storage	50	[7]
7. Hospitals	Wards and rooms	40	1,000[2]
8. Libraries	Reading rooms	60	1,000[2]
	Stack rooms	125	1,500[2]
9. Manufacturing	Light	75	2,000[2]
	Heavy	125	3,000[2]
10. Offices		50	2,000[2]
11. Printing plants	Press rooms	150	2,500[2]
	Composing and linotype rooms	100	2,000[2]
12. Residential[8]	Basic floor area	40	0[6]
	Exterior balconies	60[4]	0
	Decks	40[4]	0
13. Restrooms[9]			
14. Reviewing stands, grandstands, bleachers, and folding and telescoping seating		100	0
15. Roof decks	Same as area served or for the type of occupancy accommodated		
16. Schools	Classrooms	40	1,000[2]
17. Sidewalks and driveways	Public access	250	[7]
18. Storage	Light	125	
	Heavy	250	
19. Stores		100	3,000[2]
20. Pedestrian bridges and walkways		100	

[1]See Section 1606 for live load reductions.
[2]See Section 1604.3, first paragraph, for area of load application.
[3]Assembly areas include such occupancies as dance halls, drill rooms, gymnasiums, playgrounds, plazas, terraces and similar occupancies which are generally accessible to the public.
[4]When snow loads occur that are in excess of the design conditions, the structure shall be designed to support the loads due to the increased loads caused by drift buildup or a greater snow design as determined by the building official. See Section 1605.4. For special-purpose roofs, see Section 1605.5.
[5]Exit facilities shall include such uses as corridors serving an occupant load of 10 or more persons, exterior exit balconies, stairways, fire escapes and similar uses.
[6]Individual stair treads shall be designed to support a 300-pound (1.33 kN) concentrated load placed in a position which would cause maximum stress. Stair stringers may be designed for the uniform load set forth in the table.
[7]See Section 1604.3, second paragraph, for concentrated loads. See Table 16-B for vehicle barriers.
[8]Residential occupancies include private dwellings, apartments and hotel guest rooms.
[9]Restroom loads shall not be less than the load for the occupancy with which they are associated, but need not exceed 50 pounds per square foot (2.4 kN/m^2).

Source: Uniform Building Code. 1994 ed. Reproduced with permission of the publisher, International Conference of Building Officials.

given in the *Uniform Building Code* for determining the reduction permitted for beams, trusses, or columns that support large floor areas.

Except for floors in places of assembly (theaters, for example), and except for live loads greater than 100 psf [4.79 kN/m²], the design live load on a member may be reduced in accordance with the formula

$$R = 0.08\,(A - 150)$$

$$[R = 0.86\,(A - 14)]$$

The reduction shall not exceed 40% for horizontal members or for vertical members receiving load from one level only, 60% for other vertical members, nor R as determined by the formula

$$R = 23.1 \left(1 + \frac{D}{L} \right)$$

In these formulas

R = reduction in percent
A = area of floor supported by a member in ft²
D = unit dead load/sq ft of supported area
L = unit live load/sq ft of supported area

In office buildings and certain other building types, partitions may not be permanently fixed in location but may be erected or moved from one position to another in accordance with the requirements of the occupants. In order to provide for this flexibility, it is customary to require an allowance of 15–20 psf [0.72–0.96 kN/m²], which is usually added to other dead loads.

A.3 LATERAL LOADS

Lateral load usually refers to the effects of wind and earthquakes, which induce horizontal forces on stationary structures. Design criteria and methods in this area are continuously refined; recommended practices are presented in the various model building codes, such as the *Uniform Building Code*.

Space limitations do not permit a complete discussion of lateral loads and design for their resistance. Examples that show these criteria are in the chapters that contain examples of building structural design. For a more extensive discussion, refer to Ref. 18.

APPENDIX B

DESIGN AIDS AND DATA

This appendix provides some useful materials for rapid approximate design work. The information is grouped as follows:

General: Information that is independent of material use, relating to fundamental geometry, mechanics, and so on.

Material-Specific: Information that is specific to the use of one of the major structural materials: wood, steel, concrete, or masonry.

Foundations: Information relating to simple, bearing-type footings of sitecast reinforced concrete.

Some of this data has been adapted or directly reproduced from other publications, in which case the publishers have given permission for its use and the source publications are acknowledge.

B.1 GENERAL

Table B.1 provides useful properties of ordinary plane geometric shapes. Table B.2 provides values relating to beam structural behavior, including the finding of reactions, shears, bending moments, and dimensions of maximum deflection.

TABLE B.1   Properties of Simple Geometric Shapes

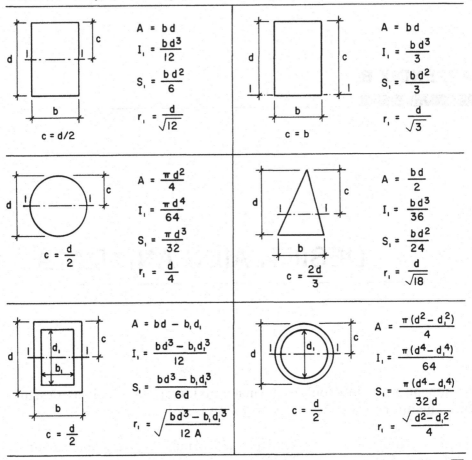

$A = \text{Area}$ $I = \text{Moment of inertia}$ $S = \text{Section modulus} = \dfrac{I}{c}$ $r = \text{Radius of gyration} = \sqrt{\dfrac{I}{A}}$

B.2   WOOD STRUCTURES

Table B.3 provides properties of structural lumber. This is an abbreviated table, limited to the more commonly used and available sizes.

Table B.4 provides values for allowable stress and modulus of elasticity for a single species of structural wood: Douglas Fir–Larch. This data is from current industry references, but you should be careful to use values as prescribed by the building code in force.

Figure B.1 provides approximate usable axial compression loads for square column shapes of various unbraced height. Note that the graphs use a specific set of data in terms of design properties; values obtained from the graphs should be used only for approximate design, with specific load capacities determined from applicable code requirements after column size and lumber choice are determined.

TABLE B.2 Values for Typical Beam Loadings

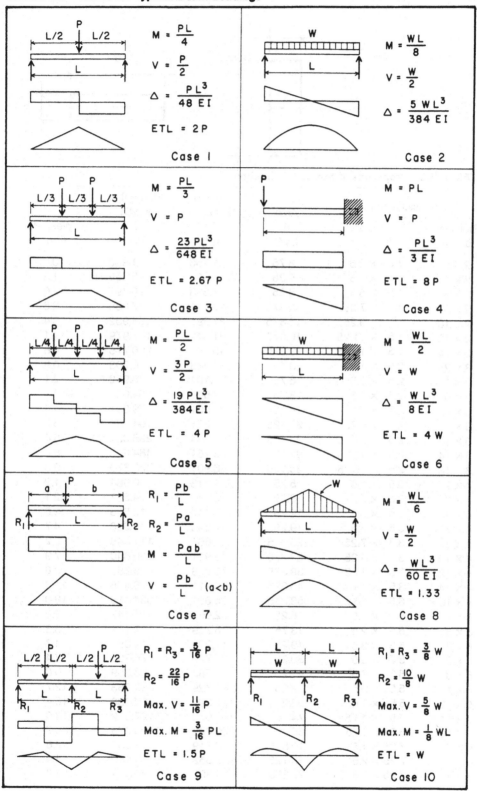

Case 1

$$M = \frac{PL}{4}$$

$$V = \frac{P}{2}$$

$$\Delta = \frac{PL^3}{48\,EI}$$

ETL = 2P

Case 2

$$M = \frac{WL}{8}$$

$$V = \frac{W}{2}$$

$$\Delta = \frac{5\,WL^3}{384\,EI}$$

Case 3

$$M = \frac{PL}{3}$$

$$V = P$$

$$\Delta = \frac{23\,PL^3}{648\,EI}$$

ETL = 2.67 P

Case 4

$$M = PL$$

$$V = P$$

$$\Delta = \frac{PL^3}{3\,EI}$$

ETL = 8P

Case 5

$$M = \frac{PL}{2}$$

$$V = \frac{3P}{2}$$

$$\Delta = \frac{19\,PL^3}{384\,EI}$$

ETL = 4P

Case 6

$$M = \frac{WL}{2}$$

$$V = W$$

$$\Delta = \frac{WL^3}{8\,EI}$$

ETL = 4W

Case 7

$$R_1 = \frac{Pb}{L}$$

$$R_2 = \frac{Pa}{L}$$

$$M = \frac{Pab}{L}$$

$$V = \frac{Pb}{L} \quad (a<b)$$

Case 8

$$M = \frac{WL}{6}$$

$$V = \frac{W}{2}$$

$$\Delta = \frac{WL^3}{60\,EI}$$

ETL = 1.33

Case 9

$$R_1 = R_3 = \frac{5}{16}P$$

$$R_2 = \frac{22}{16}P$$

$$\text{Max. } V = \frac{11}{16}P$$

$$\text{Max. } M = \frac{3}{16}PL$$

ETL = 1.5P

Case 10

$$R_1 = R_3 = \frac{3}{8}W$$

$$R_2 = \frac{10}{8}W$$

$$\text{Max. } V = \frac{5}{8}W$$

$$\text{Max. } M = \frac{1}{8}WL$$

ETL = W

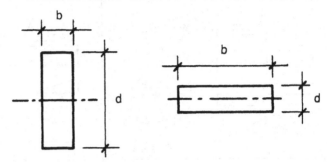

TABLE B.3 Properties of Structural Lumber

Dimensions (in.)				Area A (in.^{2})	Section Modulus S (in.^{3})	Moment of Inertia I (in.^{4})	Weight[a] (lb/ft)
Nominal		Actual					
b	h	b	h				
2 × 3		1.5	× 2.5	3.75	1.563	1.953	0.9
2 × 4		1.5	× 3.5	5.25	3.063	5.359	1.3
2 × 6		1.5	× 5.5	8.25	7.563	20.797	2.0
2 × 8		1.5	× 7.25	10.875	13.141	47.635	2.6
2 × 10		1.5	× 9.25	13.875	21.391	98.932	3.4
2 × 12		1.5	× 11.25	16.875	31.641	177.979	4.1
2 × 14		1.5	× 13.25	19.875	43.891	290.775	4.8
3 × 2		2.5	× 1.5	3.75	0.938	0.703	0.9
3 × 4		2.5	× 3.5	8.75	5.104	8.932	2.1
3 × 6		2.5	× 5.5	13.75	12.604	34.661	3.3
3 × 8		2.5	× 7.25	18.125	21.901	79.391	4.4
3 × 10		2.5	× 9.25	23.125	35.651	164.886	5.6
3 × 12		2.5	× 11.25	28.125	52.734	296.631	6.8
3 × 14		2.5	× 13.25	33.125	73.151	484.625	8.1
3 × 16		2.5	× 15.25	38.125	96.901	738.870	9.3
4 × 2		3.5	× 1.5	5.25	1.313	0.984	1.3
4 × 3		3.5	× 2.5	8.75	3.646	4.557	2.1
4 × 4		3.5	× 3.5	12.25	7.146	12.505	3.0
4 × 6		3.5	× 5.5	19.25	17.646	48.526	4.7
4 × 8		3.5	× 7.25	23.375	30.661	111.148	6.2
4 × 10		3.5	× 9.25	32.375	49.911	230.840	7.9
4 × 12		3.5	× 11.25	39.375	73.828	415.283	9.6
4 × 14		3.5	× 13.25	46.375	102.411	678.475	11.3
4 × 16		3.5	× 15.25	53.375	135.661	1034.418	13.0
6 × 2		5.5	× 1.5	8.25	2.063	1.547	2.0
6 × 3		5.5	× 2.5	13.75	5.729	7.161	3.3
6 × 4		5.5	× 3.5	19.25	11.229	19.651	4.7
6 × 6		5.5	× 5.5	30.25	27.729	76.255	7.4
6 × 8		5.5	× 7.5	41.25	51.563	193.359	10.0
6 × 10		5.5	× 9.5	52.25	82.729	392.963	12.7
6 × 12		5.5	× 11.5	63.25	121.229	697.068	15.4
6 × 14		5.5	× 13.5	74.25	167.063	1127.672	18.0
6 × 16		5.5	× 15.5	85.25	220.229	1706.776	20.7
8 × 2		7.25	× 1.5	10.875	2.719	2.039	2.6
8 × 3		7.25	× 2.5	18.125	7.552	9.440	4.4
8 × 4		7.25	× 3.5	25.375	14.802	25.904	6.2

TABLE B.3 *(Continued)*

Dimensions (in.)				Area	Section Modulus	Moment of Inertia	
Nominal		Actual		A	S	I	Weight[a]
b	h	b	h	(in.^{2})	(in.^{3})	(in.^{4})	(lb/ft)
8 × 6		7.5	× 5.5	41.25	37.813	103.984	10.0
8 × 8		7.5	× 7.5	56.25	70.313	263.672	13.7
8 × 10		7.5	× 9.5	71.25	112.813	535.859	17.3
8 × 12		7.5	× 11.5	86.25	165.313	950.547	21.0
8 × 14		7.5	× 13.5	101.25	227.813	1537.734	24.6
8 × 16		7.5	× 15.5	116.25	300.313	2327.422	28.3
8 × 18		7.5	× 17.5	131.25	382.813	3349.609	31.9
8 × 20		7.5	× 19.5	146.25	475.313	4634.297	35.5
10 × 10		9.5	× 9.5	90.25	142.896	678.755	21.9
10 × 12		9.5	× 11.5	109.25	209.396	1204.026	26.6
10 × 14		9.5	× 13.5	128.25	288.563	1947.797	31.2
10 × 16		9.5	× 15.5	147.25	380.396	2948.068	35.8
10 × 18		9.5	× 17.5	166.25	484.896	4242.836	40.4
10 × 20		9.5	× 19.5	185.25	602.063	5870.109	45.0
12 × 12		11.5	× 11.5	132.25	253.479	1457.505	32.1
12 × 14		11.5	× 13.5	155.25	349.313	2357.859	37.7
12 × 16		11.5	× 15.5	178.25	460.479	3568.713	43.3
12 × 18		11.5	× 17.5	201.25	586.979	5136.066	48.9
12 × 20		11.5	× 19.5	224.25	728.813	7105.922	54.5
12 × 22		11.5	× 21.5	247.25	885.979	9524.273	60.1
12 × 24		11.5	× 23.5	270.25	1058.479	12437.129	65.7
14 × 14		13.5	× 13.5	182.25	410.063	2767.922	44.3
16 × 16		15.5	× 15.5	240.25	620.646	4810.004	58.4

[a] Based on an assumed average weight of 35 lb/ft^3.

Source: Compiled from data in the *National Design Specification for Wood Construction*, 1991 ed. (Ref. 4), with permission of the publisher, National Forest Products Association.

Figure B.2 provides data relating to design for beam deflection. You can use graphs to determine approximate deflections, but note that specific values for maximum bending stress and modulus of elasticity have been used to plot the graphs. Actual deflections may be determined by direct proportion if other values for stress and modulus of elasticity are known. The graph is more useful, however, during early design work, enabling you to quickly evaluate the relative degree of concern for deflection: the lines representing specific span/deflection ratios indicate how deflection relates to specific beam depth. In many cases you may discover that deflection will not be critical and thus the design can be focused on other issues.

Tables B.5 and B.6 provide data for load capacities of structural plywood decking, as presented in the 1994 UBC (Ref. 1). Note the relation of the grain direction of the plywood face plies to the direction of the support members for the deck.

TABLE B.4 Design Values for Visually-Graded Lumber of Douglas Fir–Larch[a] (Values in psi)

Species and Commercial Grade	Size and Use Classification	Extreme Fiber in Bending F_b		Tension Parallel to Grain F_t	Horizontal Shear F_v	Compression Perpendicular to Grain $F_{c\perp}$	Compression Parallel to Grain F_c	Modulus of Elasticity E
		Single Member Uses	Repetitive Member Uses					
Dimension Lumber (Moisture not exceeding 19%)								
Select Structural	2 to 4 in. thick,	1450	1668	1000	95	625	1700	1,900,000
No. 1 and better	2 in. and wider	1150	1323	775	95	625	1500	1,500,000
No. 1		1000	1150	675	95	625	1450	1,700,000
No. 2		875	1006	575	95	625	1300	1,600,000
No. 3		500	575	325	95	625	750	1,400,000
Stud		675	776	450	95	625	525	1,400,000
Construction		1000	1150	650	95	625	1600	1,500,000
Standard		550	663	375	95	625	1350	1,400,000
Utility		275	316	175	95	625	575	1,300,000
Timbers (Surfaced Green)								
Dense sel. struc.	Beams and	1900	—	1100	85	730	1300	1,700,000
Select Structural	Stringers	1600	—	950	85	625	1100	1,600,000
Dense No. 1		1550	—	775	85	730	1100	1,700,000
No. 1		1300	—	675	85	625	925	1,600,000
No. 2		875	—	425	85	625	600	1,300,000
Dense sel. struc.	Posts and	1750	—	1150	85	730	1350	1,700,000
Select structural	Timbers	1500	—	1000	85	625	1150	1,600,000
Dense No. 1		1400	—	950	85	730	1200	1,700,000
No. 1		1200	—	525	85	625	1000	1,600,000
No. 2		750	—	475	85	625	700	1,300,000
Decking (Moisture not to exceed 19%)								
Select dex	Decking	1750	2013	—	—	625	—	1,500,000
Commercial dex		1450	1668	—	—	625	—	1,500,000

[a] Values listed are for normal duration loading with wood that is surfaced dry or green and used at 19% maximum moisture content.

Source: Data adapted from *National Design Specification for Wood Construction*, 1991 ed. (Ref. 4), with permission of the publisher, National Forest Products Association. The table in the reference document lists several other wood species and has extensive footnotes.

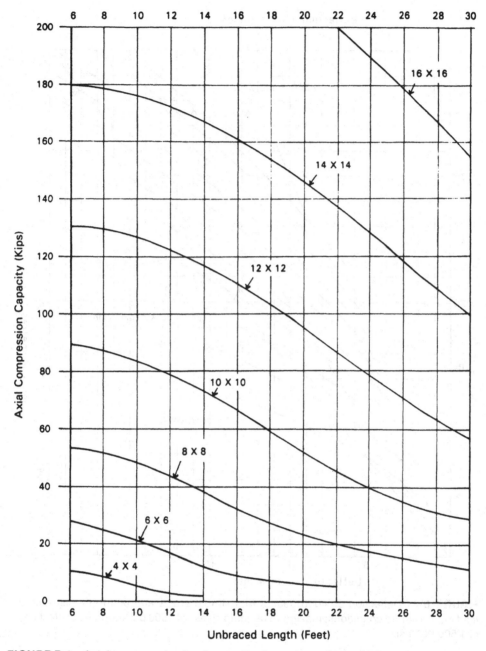

FIGURE B.1 Axial compression load capacity of wood members with square cross sections. Based on normal loading conditions for Douglas Fir–Larch, No. 1 grade.

Table B.7 provides values for shear capacity of plywood walls used as vertical diaphragms. There are many variables incorporated in this table (taken from the 1994 UBC–Ref. 1). Readers unfamiliar with diaphragm design should refer to more extensive discussions in various references, such as *Simplified Building Design for Wind and Earthquake Forces* (Ref. 18). Although

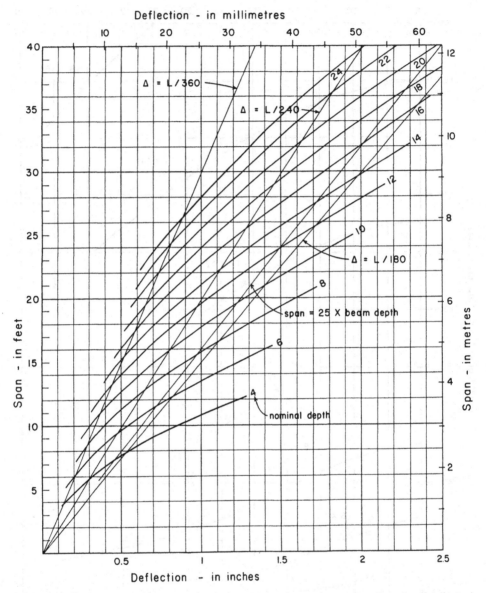

FIGURE B.2 Deflection of wood beams with simple spans and uniformly distributed loading. Assumed conditions: maximum bending stress of 1500 psi, modulus of elasticity of 1,500,000 psi.

the information in Table B.7 is published in the UBC, it actually comes from the plywood industry, which sponsors the studies and testing used to establish the table data.

Similar in some ways to Table B.7, Table B.8 provides data for plywood horizontal diaphragms.

Tables B.9 through B.14 provide data for light wood frame members for horizontal spanning functions. This data is reproduced directly from the

TABLE B.5 Data for Plywood Decks—Continuous Over Two or More Supports and Face Grain Perpendicular to Supports[1,2] (UBC Table 23-I-S-1)

SHEATHING GRADES

Panel Span Rating	Panel Thickness (inches)	ROOF[3] Maximum Span (inches) × 25.4 for mm		Load[5] (pounds per square foot) × 0.0479 for kN/m²		FLOOR[4] Maximum Span (inches)
Roof/Floor Span	× 25.4 for mm	With Edge Support[6]	Without Edge Support	Total Load	Live Load	× 25.4 for mm
12/0	5/16	12	12	40	30	0
16/0	5/16, 3/8	16	16	40	30	0
20/0	5/16, 3/8	20	20	40	30	0
24/0	3/8, 7/16, 1/2	24	20[7]	40	30	0
24/16	7/16, 1/2	24	24	50	40	16
32/16	15/32, 1/2, 5/8	32	28	40	30	16[8]
40/20	19/32, 5/8, 3/4, 7/8	40	32	40	30	20[8,9]
48/24	23/32, 3/4, 7/8	48	36	45	35	24
54/32	7/8, 1	54	40	45	35	32
60/48	7/8, 1, 1 1/8	60	48	45	35	48

SINGLE-FLOOR GRADES

Panel Span Rating (inches)	Panel Thickness (inches)	ROOF[3] Maximum Span (inches) × 25.4 for mm		Load[5] (pounds per square foot) × 0.0479 for kN/m²		FLOOR[4] Maximum Span (inches)
	× 25.4 for mm	With Edge Support[6]	Without Edge Support	Total Load	Live Load	× 25.4 for mm
16 oc	1/2, 19/32, 5/8	24	24	50	40	16[8]
20 oc	19/32, 5/8, 3/4	32	32	40	30	20[8,9]
24 oc	23/32, 3/4	48	36	35	25	24
32 oc	7/8, 1	48	40	50	40	32
48 oc	1 3/32, 1 1/8	60	48	50	50	48

[1] Applies to panels 24 inches (610 mm) or wider.
[2] Floor and roof sheathing conforming with this table shall be deemed to meet the design criteria of Section 2321.
[3] Uniform load deflection limitations: $1/180$ of span under live load plus dead load. $1/240$ under live load only.
[4] Panel edges shall have approved tongue-and-groove joints or shall be supported with blocking unless $1/4$-inch (6.4 mm) minimum thickness underlayment or $1\,1/2$ inches (38 mm) of approved cellular or lightweight concrete is placed over the subfloor, or finish floor is $3/4$-inch (19 mm) wood strip. Allowable uniform load based on deflection of $1/360$ of span is 100 pounds per square foot (psf) (4.79 kN/m²) except the span rating of 48 inches on center is based on a total load of 65 psf (3.11 kN/m).
[5] Allowable load at maximum span.
[6] Tongue-and-groove edges, panel edge clips [one midway between each support, except two equally spaced between supports 48 inches (1219 mm) on center], lumber blocking, or other. Only lumber blocking shall satisfy blocked diaphragms requirements.
[7] For $1/2$-inch (13 mm) panel, maximum span shall be 24 inches (610 mm).
[8] May be 24 inches (610 mm) on center where $3/4$-inch (19 mm) wood strip flooring is installed at right angles to joist.
[9] May be 24 inches (610 mm) on center for floors where $1\,1/2$ inches (38 mm) of cellular or lightweight concrete is applied over the panels.

Source: Uniform Building Code, 1994 ed. Reproduced with permission of the publisher, International Conference of Building Officials.

TABLE B.6 Data for Plywood Decks—Continuous Over Two or More Supports and Face Grain Parallel to Supports[1,2] (UBC Table 23-I-S-2)

PANEL GRADE	THICKNESS (inch) × 25.4 for mm	MAXIMUM SPAN (inches)	LOAD AT MAXIMUM SPAN (psf) × 0.0479 for kN/m²	
			Live	Total
Structural I	7/16	24	20	30
	15/32	24	35³	45³
	1/2	24	40³	50³
	19/32, 5/8	24	70	80
	23/32, 3/4	24	90	100
Other grades covered in U.B.C. Standard 23-2 or 23-3	7/16	16	40	50
	15/32	24	20	25
	1/2	24	25	30
	19/32	24	40³	50³
	5/8	24	45³	55³
	23/32, 3/4	24	60³	65³

[1]Roof sheathing conforming with this table shall be deemed to meet the design criteria of Section 2321.

[2]Uniform load deflection limitations: $1/180$ of span under live load plus dead load, $1/240$ under live load only. Edges shall be blocked with lumber or other approved type of edge supports.

[3]For composite and four-ply plywood structural panel, load shall be reduced by 15 pounds per square foot (0.72 kN/m²).

Source: Uniform Building Code, 1994 ed. Reproduced with permission of the publisher, International Conference of Building Officials.

TABLE B.7 Allowable Shear in Pounds per Foot for Shear Walls of Structural Plywood Nailed to Framing of Douglas Fir–Larch or Southern Pine[1,2] (UBC Table 23-I-K-1)

PANEL GRADE	MINIMUM NOMINAL PANEL THICKNESS (inches) ×25.4 for mm	MINIMUM NAIL PENETRATION IN FRAMING (inches) ×25.4 for mm	PANELS APPLIED DIRECTLY TO FRAMING — Nail Size (Common or Galvanized Box)	Nail Spacing at Panel Edges (in.) ×25.4 for mm — 6 (×0.0146 for N/mm)	4	3	2³	PANELS APPLIED OVER ¹⁄₂-INCH (13 mm) OR ⁵⁄₈-INCH (16 mm) GYPSUM SHEATHING — Nail Size (Common or Galvanized Box)	Nail Spacing at Panel Edges (in.) ×25.4 for mm — 6 (×0.0146 for N/mm)	4	3	2³
Structural I	5/16	1 1/4	6d	200	300	390	510	8d	200	300	390	510
	3/8	1 1/2	8d	230[4]	360[4]	460[4]	610[4]	10d[5]	280	430	550	730
	7/16		8d	255[4]	395[4]	505[4]	670[4]					
	15/32		8d	280	430	550	730					
	15/32	1 5/8	10d[5]	340	510	665	870	—	—	—	—	—
C-D, C-C Sheathing, plywood panel siding and other grades covered in U.B.C. Standard 23-2 or 23-3	5/16	1 1/4	6d	180	270	350	450	8d	180	270	350	450
	3/8		6d	200	300	390	510	8d	200	300	390	510
	3/8	1 1/2	8d	220[4]	320[4]	410[4]	530[4]					
	7/16		8d	240[4]	350[4]	450[4]	585[4]	10d[5]	260	380	490	640
	15/32		8d	260	380	490	640					
	15/32	1 5/8	10d[5]	310	460	600	770					
	19/32		10d[5]	340	510	665	870	—	—	—	—	—
			Nail Size (Galvanized Casing)					**Nail Size (Galvanized Casing)**				
Plywood panel siding in grades covered in U.B.C. Standard 23-2	5/16	1 1/4	6d	140	210	275	360	8d	140	210	275	360
	3/8	1 1/2	8d	160	240	310	410	10d[5]	160	240	310	410

[1]All panel edges backed with 2-inch (51 mm) nominal or wider framing. Panels installed either horizontally or vertically. Space nails at 6 inches (152 mm) on center along intermediate framing members for 3/8-inch (9.5 mm) and 7/16-inch (11 mm) panels installed on studs spaced 24 inches (610 mm) on center and 12 inches (305 mm) on center for other conditions and panel thicknesses. These values are for short-time loads due to wind or earthquake and must be reduced 25 percent for normal loading. Allowable shear values for nails in framing members of other species set forth in Table 23-III-FF of Division III shall be calculated for all other grades by multiplying the shear capacities for nails in Structural I by the following factors: 0.82 for species with specific gravity greater than or equal to 0.42 but less than 0.49, and 0.65 for species with a specific gravity less than 0.42.

[2]Where panels are applied on both faces of a wall and nail spacing is less than 6 inches (152 mm) on center on either side, panel joints shall be offset to fall on different framing members or framing shall be 3-inch (76 mm) nominal or thicker and nails on each side shall be staggered.

[3]Framing at adjoining panel edges shall be 3-inch (76 mm) nominal or wider and nails shall be staggered where nails are spaced 2 inches (51 mm) on center.

TABLE B.8 Allowable Shear in Pounds per Foot for Horizontal Diaphragms of Structural Plywood Nailed to Framing of Douglas Fir–Larch or Southern Pine[1] (UBC Table 23-I-J-1)

PANEL GRADE	COMMON NAIL SIZE	MINIMUM NAIL PENETRATION IN FRAMING (inches)	MINIMUM NOMINAL PANEL THICKNESS (inches) × 25.4 for mm	MINIMUM NOMINAL WIDTH OF FRAMING MEMBER (inches)	BLOCKED DIAPHRAGMS — Nail spacing (in.) at diaphragm boundaries (all cases), at continuous panel edges parallel to load (Cases 3 and 4) and at all panel edges (Cases 5 and 6) × 25.4 for mm				UNBLOCKED DIAPHRAGMS — Nails spaced 6" (152 mm) max. at supported edges	
					6	4	2½[2]	2[2]	Case 1 (No unblocked edges or continuous joints parallel to load)	All other configurations (Cases 2, 3, 4, 5 and 6)
					Nail spacing (in.) at other panel edges × 25.4 for mm					
					6	6	4	3		
					× 0.0146 for N/mm					
Structural 1	6d	1¼	5/16	2	185	250	375	420	165	125
				3	210	280	420	475	185	140
	8d	1½	3/8	2	270	360	530	600	240	180
				3	300	400	600	675	265	200
	10d[3]	1⅝	15/32	2	320	425	640	730	285	215
				3	360	480	720	820	320	240
C-D, C-C, Sheathing, and other grades covered in U.B.C. Standard 23-3 or 23-9	6d	1¼	5/16	2	170	225	335	380	150	110
				3	190	250	380	430	170	125
			3/8	2	185	250	375	420	165	125
				3	210	280	420	475	185	140
	8d	1½	3/8	2	240	320	480	545	215	160
				3	270	360	540	610	240	180
			7/16	2	255	340	505	575	230	170
				3	285	380	570	645	255	190
			15/32	2	270	360	530	600	240	180
				3	300	400	600	675	265	200
	10d[3]	1⅝	15/32	2	290	385	575	655	255	190
				3	325	430	650	735	290	215
			19/32	2	320	425	640	730	285	215
				3	360	480	720	820	320	240

[1] These values are for short-time loads due to wind or earthquake and must be reduced 25 percent for normal loading. Space nails 12 inches (305 mm) on center along intermediate framing members.

Allowable shear values for nails in framing members of other species set forth in Table 23-III-FF of Division III shall be calculated for all other grades by multiplying the shear capacities for nails in Structural I by the following factors: 0.82 for species with specific gravity greater than or equal to 0.42 but less than 0.49, and 0.65 for species with a specific gravity less than 0.42.

[2] Framing at adjoining panel edges shall be 3-inch (76 mm) nominal or wider and nails shall be staggered where nails are spaced 2 inches (51 mm) or 2 1/2 inches (64 mm) on center.

[3] Framing at adjoining panel edges shall be 3-inch (76 mm) nominal or wider and nails shall be staggered where 10d nails having penetration into framing of more than 1 5/8 inches (41 mm) are spaced 3 inches (76 mm) or less on center.

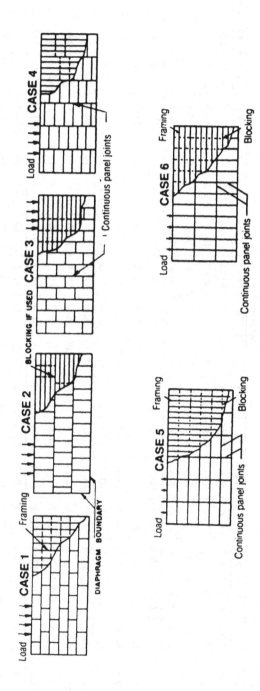

NOTE:: Framing may be oriented in either direction for diaphragms, provided sheathing is properly designed for vertical loading.

TABLE B.9 Allowable Spans in Feet for Floor Joists—Live Load 40 psf, Dead Load 20 psf, Live Load Deflection L/360 (UBC Table 23-I-V-J-2)

DESIGN CRITERIA:
Deflection — For 40 (1.92 kN/m²) psf live load.
Limited to span in inches (mm) divided by 360.
Strength — Live load of 40 psf (1.92 kN/m²) plus dead load of 20 psf (0.96 kN/m²) determines the required bending design value.

Joist Size (in) × 25.4 for mm	Spacing (in) × 25.4 for mm	Modulus of Elasticity, E, in 1,000,000 psi × 0.00689 for N/mm²																
		0.8	0.9	1.0	1.1	1.2	1.3	1.4	1.5	1.6	1.7	1.8	1.9	2.0	2.1	2.2	2.3	2.4
2 × 6	12.0	8-6	8-10	9-2	9-6	9-9	10-0	10-3	10-6	10-9	10-11	11-2	11-4	11-7	11-9	11-11	12-1	12-3
	16.0	7-9	8-0	8-4	8-7	8-10	9-1	9-4	9-6	9-9	9-11	10-2	10-4	10-6	10-8	10-10	11-0	11-2
	19.2	7-3	7-7	7-10	8-1	8-4	8-7	8-9	9-0	9-2	9-4	9-6	9-8	9-10	10-0	10-2	10-4	10-6
	24.0	6-9	7-0	7-3	7-6	7-9	7-11	8-2	8-4	8-6	8-8	8-10	9-0	9-2	9-4	9-6	9-7	9-9
2 × 8	12.0	11-3	11-8	12-1	12-6	12-10	13-2	13-6	13-10	14-2	14-5	14-8	15-0	15-3	15-6	15-9	15-11	16-2
	16.0	10-2	10-7	11-0	11-4	11-8	12-0	12-3	12-7	12-10	13-1	13-4	13-7	13-10	14-1	14-3	14-6	14-8
	19.2	9-7	10-0	10-4	10-8	11-0	11-3	11-7	11-10	12-1	12-4	12-7	12-10	13-0	13-3	13-5	13-8	13-10
	24.0	8-11	9-3	9-7	9-11	10-2	10-6	10-9	11-0	11-3	11-5	11-8	11-11	12-1	12-3	12-6	12-8	12-10
2 × 10	12.0	14-4	14-11	15-5	15-11	16-5	16-10	17-3	17-8	18-0	18-5	18-9	19-1	19-5	19-9	20-1	20-4	20-8
	16.0	13-0	13-6	14-0	14-6	14-11	15-3	15-8	16-0	16-5	16-9	17-0	17-4	17-8	17-11	18-3	18-6	18-9
	19.2	12-3	12-9	13-2	13-7	14-0	14-5	14-9	15-1	15-5	15-9	16-0	16-4	16-7	16-11	17-2	17-5	17-8
	24.0	11-4	11-10	12-3	12-8	13-0	13-4	13-8	14-0	14-4	14-7	14-11	15-2	15-5	15-8	15-11	16-2	16-5
2 × 12	12.0	17-5	18-1	18-9	19-4	19-11	20-6	21-0	21-6	21-11	22-5	22-10	23-3	23-7	24-0	24-5	24-9	25-1
	16.0	15-10	16-5	17-0	17-7	18-1	18-7	19-1	19-6	19-11	20-4	20-9	21-1	21-6	21-10	22-2	22-6	22-10
	19.2	14-11	15-6	16-0	16-7	17-0	17-6	17-11	18-4	18-9	19-2	19-6	19-10	20-2	20-6	20-10	21-2	21-6
	24.0	13-10	14-4	14-11	15-4	15-10	16-3	16-8	17-0	17-5	17-9	18-1	18-5	18-9	19-1	19-4	19-8	19-11
F_b	12.0	862	932	1,000	1,066	1,129	1,191	1,251	1,310	1,368	1,424	1,480	1,534	1,587	1,640	1,692	1,742	1,793
	16.0	949	1,026	1,101	1,173	1,243	1,311	1,377	1,442	1,506	1,568	1,629	1,688	1,747	1,805	1,862	1,918	1,973
	19.2	1,008	1,090	1,170	1,246	1,321	1,393	1,464	1,533	1,600	1,666	1,731	1,794	1,857	1,918	1,978	2,038	2,097
	24.0	1,086	1,174	1,260	1,343	1,423	1,501	1,577	1,651	1,724	1,795	1,864	1,933	2,000	2,066	2,131	2,195	2,258

NOTE: The required bending design value, F_b, in pounds per square inch ($\times$ 0.00689 for N/mm²) is shown at the bottom of this table and is applicable to all lumber sizes shown. Spans are shown in feet-inches (1 foot = 304.8 mm, 1 inch = 25.4 mm) and are limited to 26 feet (7925 mm) and less.

TABLE B.10 Allowable Spans in Feet for Ceiling Joists—Live Load 20 psf, Dead Load 10 psf, Live Load Deflection L/240 (UBC Table 23-I-V-J-4)

DESIGN CRITERIA:
Deflection — For 20 psf (0.96 kN/m²) live load.
Limited to span in inches (mm) divided by 240.
Strength — Live load of 20 psf (0.96 kN/m²) plus dead load of 10 psf (0.48 kN/m²) determines the required bending design value.

Joist Size (in)	Spacing (in)	Modulus of Elasticity, E, in 1,000,000 psi × 0.00689 for N/mm²																
× 25.4 for mm	× 25.4 for mm	0.8	0.9	1.0	1.1	1.2	1.3	1.4	1.5	1.6	1.7	1.8	1.9	2.0	2.1	2.2	2.3	2.4
2 × 4	12.0	7-10	8-1	8-5	8-8	8-11	9-2	9-5	9-8	9-10	10-0	10-3	10-5	10-7	10-9	10-11	11-1	11-3
	16.0	7-1	7-5	7-8	7-11	8-1	8-4	8-7	8-9	8-11	9-1	9-4	9-6	9-8	9-9	9-11	10-1	10-3
	19.2	6-8	6-11	7-2	7-5	7-8	7-10	8-1	8-3	8-5	8-7	8-9	8-11	9-1	9-3	9-4	9-6	9-8
	24.0	6-2	6-5	6-8	6-11	7-1	7-3	7-6	7-8	7-10	8-0	8-1	8-3	8-5	8-7	8-8	8-10	8-11
2 × 6	12.0	12-3	12-9	13-3	13-8	14-1	14-5	14-9	15-2	15-6	15-9	16-1	16-4	16-8	16-11	17-2	17-5	17-8
	16.0	11-2	11-7	12-0	12-5	12-9	13-1	13-5	13-9	14-1	14-4	14-7	14-11	15-2	15-5	15-7	15-10	16-1
	19.2	10-6	10-11	11-4	11-8	12-0	12-4	12-8	12-11	13-3	13-6	13-9	14-0	14-3	14-6	14-8	14-11	15-2
	24.0	9-9	10-2	10-6	10-10	11-2	11-5	11-9	12-0	12-3	12-6	12-9	13-0	13-3	13-5	13-8	13-10	14-1
2 × 8	12.0	16-2	16-10	17-5	18-0	18-6	19-0	19-6	19-11	20-5	20-10	21-2	21-7	21-11	22-4	22-8	23-0	23-4
	16.0	14-8	15-3	15-10	16-4	16-10	17-3	17-9	18-1	18-6	18-11	19-3	19-7	19-11	20-3	20-7	20-11	21-2
	19.2	13-10	14-5	14-11	15-5	15-10	16-3	16-8	17-1	17-5	17-9	18-1	18-5	18-9	19-1	19-5	19-8	19-11
	24.0	12-10	13-4	13-10	14-3	14-8	15-1	15-6	15-10	16-2	16-6	16-10	17-2	17-5	17-9	18-0	18-3	18-6
2 × 10	12.0	20-8	21-6	22-3	22-11	23-8	24-3	24-10	25-5	26-0								
	16.0	18-9	19-6	20-2	20-10	21-6	22-1	22-7	23-1	23-8	24-1	24-7	25-0	25-5	25-10			
	19.2	17-8	18-4	19-0	19-7	20-2	20-9	21-3	21-9	22-3	22-8	23-1	23-7	23-11	24-4	24-9	25-1	25-5
	24.0	16-5	17-0	17-8	18-3	18-9	19-3	19-9	20-2	20-8	21-1	21-6	21-10	22-3	22-7	22-11	23-4	23-8
F_b	12.0	896	969	1,040	1,108	1,174	1,239	1,302	1,363	1,423	1,481	1,539	1,595	1,651	1,706	1,759	1,812	1,864
	16.0	986	1,067	1,145	1,220	1,293	1,364	1,433	1,500	1,566	1,631	1,694	1,756	1,817	1,877	1,936	1,995	2,052
	19.2	1,048	1,134	1,216	1,296	1,374	1,449	1,522	1,594	1,664	1,733	1,800	1,866	1,931	1,995	2,058	2,120	2,181
	24.0	1,129	1,221	1,310	1,396	1,480	1,561	1,640	1,717	1,793	1,866	1,939	2,010	2,080	2,149	2,217	2,283	2,349

NOTE: The required bending design value, F_b, in pounds per square inch (× 0.00689 for N/mm²) is shown at the bottom of this table and is applicable to all lumber sizes shown. Spans are shown in feet-inches (1 foot = 304.8 mm, 1 inch = 25.4 mm) and are limited to 26 feet (7925 mm) and less.

TABLE B.11 Allowable Spans in Feet for Rafters—Live Load 20 psf, Dead Load 10 psf, Live Load Deflection L/240 (UBC Table 23-I-V-R-1)

DESIGN CRITERIA:

Strength — Live load of 20 psf (0.96 kN/m²) plus dead load of 10 psf (0.48 kN/m²) determines the required bending design value.

Deflection — For 20 psf (0.96 kN/m²) live load.

Limited to span in inches (mm) divided by 240.

Bending Design Value, F_b (psi) — × 0.00689 for N/mm²

Rafter Size (in) ×25.4 for mm	Spacing (in)	300	400	500	600	700	800	900	1000	1100	1200	1300	1400	1500	1600	1700	1800	1900	2000	2100	2200	2300	2400
2x6	12.0	7-1	8-2	9-2	10-0	10-10	11-7	12-4	13-0	13-7	14-2	14-8	15-4	15-11	16-5	16-11	17-5	17-10					
2x6	16.0	6-2	7-1	7-11	8-8	9-5	10-0	10-8	11-3	11-9	12-4	12-10	13-3	13-9	14-2	14-8	15-1	15-6	15-11	16-3			
2x6	19.2	5-7	6-6	7-3	7-11	8-7	9-2	9-8	10-3	10-9	11-3	11-8	12-2	12-7	13-0	13-4	13-8	14-2	14-6	14-10	15-2	15-7	
2x6	24.0	5-0	5-10	6-6	7-1	7-8	8-2	8-8	9-2	9-7	10-0	10-5	10-10	11-3	11-7	11-11	12-4	12-8	13-0	13-3	13-7	13-11	14-2
2x8	12.0	9-4	10-10	12-1	13-3	14-4	15-3	16-3	17-1	17-11	18-9	19-6	20-3	20-11	21-7	22-3	22-11	23-7					
2x8	16.0	8-1	9-4	10-6	11-6	12-5	13-3	14-0	14-10	15-6	16-3	16-10	17-6	18-1	18-8	19-4	19-10	20-5	20-11	21-5			
2x8	19.2	7-5	8-7	9-7	10-6	11-4	12-1	12-10	13-6	14-2	14-10	15-5	16-0	16-7	17-1	17-7	18-1	18-7	19-1	19-7	20-0	20-6	
2x8	24.0	6-7	7-8	8-7	9-4	10-1	10-10	11-6	12-1	12-8	13-3	13-9	14-4	14-10	15-3	15-9	16-3	16-8	17-1	17-8	17-11	18-4	18-9
2x10	12.0	11-11	13-9	15-5	16-11	18-3	19-6	20-8	21-10	22-10	23-11	24-10	25-10										
2x10	16.0	10-4	11-11	13-4	14-8	15-10	16-11	17-11	18-11	19-10	20-8	21-6	22-4	23-1	23-11	24-7	25-4	26-0					
2x10	19.2	9-5	10-11	12-2	13-4	14-5	15-5	16-4	17-3	18-1	18-11	19-8	20-5	21-1	21-10	22-6	23-1	23-9	24-5	25-0	25-7		
2x10	24.0	8-5	9-8	10-11	11-11	12-11	13-9	14-8	15-5	16-2	16-11	17-7	18-3	18-11	19-6	20-1	20-8	21-3	21-10	22-4	22-10	23-5	23-11
2x12	12.0	14-6	16-9	18-9	20-6	22-2	23-9	25-2															
2x12	16.0	12-7	14-6	16-3	17-9	19-3	20-6	21-9	23-0	24-1	25-2												
2x12	19.2	11-6	13-3	14-10	16-3	17-6	18-9	19-11	21-0	22-0	23-0	23-11	24-10	25-8									
2x12	24.0	10-3	11-10	13-3	14-6	15-8	16-9	17-9	18-9	19-8	20-6	21-5	22-2	23-0	23-9	24-5	25-2	25-10					
E	12.0	0.15	0.24	0.33	0.44	0.55	0.67	0.80	0.94	1.09	1.24	1.40	1.56	1.73	1.91	2.09	2.28	2.47					
E	16.0	0.13	0.21	0.29	0.38	0.48	0.58	0.70	0.82	0.94	1.07	1.21	1.35	1.50	1.65	1.81	1.97	2.14	2.31	2.48			
E	19.2	0.12	0.19	0.26	0.35	0.44	0.53	0.64	0.75	0.86	0.98	1.10	1.23	1.37	1.51	1.65	1.80	1.95	2.11	2.27	2.43	2.60	
E	24.0	0.11	0.17	0.24	0.31	0.39	0.48	0.57	0.67	0.77	0.88	0.99	1.10	1.22	1.35	1.48	1.61	1.75	1.89	2.03	2.18	2.33	2.48

NOTE: The required modulus of elasticity, E, in 1,000,000 pounds per square inch (psi) (× 0.00689 for N/mm²) is shown at the bottom of this table, is limited to 2.6 million psi (17 914 N/mm²) and less, and is applicable to all lumber sizes shown. Spans are shown in feet-inches (1 foot = 304.8 mm, 1 inch = 25.4 mm) and are limited to 26 feet (7925 mm) and less.

TABLE B.12 Allowable Spans in Feet for Rafters—Live Load 30 psf, Dead Load 10 psf, Live Load Deflection L/240 (UBC Table 23-I-V-R-2)

DESIGN CRITERIA:
Strength — Live load of 30 (1.44 kN/m²) psf plus dead load of 10 psf (0.48 kN/m²) determines the required bending design value.
Deflection — For 30 psf (1.44 kN/m²) live load.
Limited to span in inches (mm) divided by 240.

Rafter Size (in) × 25.4 for mm	Spacing (in) × 25.4 for mm	Bending Design Value, F_b (psi) × 0.00689 for N/mm²																					
		300	400	500	600	700	800	900	1000	1100	1200	1300	1400	1500	1600	1700	1800	1900	2000	2100	2200	2300	2400
2x6	12.0	6-2	7-1	7-11	8-8	9-5	10-0	10-8	11-3	11-9	12-4	12-10	13-3	13-9	14-2	14-8	15-1	15-6	15-11				
	16.0	5-4	6-2	6-10	7-6	8-2	8-8	9-3	9-9	10-2	10-8	11-1	11-6	11-11	12-4	12-8	13-1	13-5	13-9	14-1	14-5		
	19.2	4-10	5-7	6-3	6-10	7-5	7-11	8-5	8-11	9-4	9-9	10-1	10-6	10-10	11-3	11-7	11-11	12-3	12-7	12-10	13-2	13-6	
	24.0	4-4	5-0	5-7	6-2	6-8	7-1	7-6	7-11	8-4	8-8	9-1	9-5	9-9	10-0	10-4	10-8	10-11	11-3	11-6	11-9	12-0	12-4
2x8	12.0	8-1	9-4	10-6	11-6	12-5	13-3	14-0	14-10	15-6	16-3	16-10	17-6	18-1	18-9	19-4	19-10	20-5	20-11				
	16.0	7-0	8-1	9-1	9-11	10-9	11-6	12-2	12-10	13-5	14-0	14-7	15-2	15-8	16-3	16-9	17-2	17-8	18-1	18-7	19-0		
	19.2	6-5	7-5	8-3	9-1	9-9	10-6	11-1	11-8	12-3	12-10	13-4	13-10	14-4	14-10	15-3	15-8	16-2	16-7	16-11	17-4	17-9	
	24.0	5-8	6-7	7-5	8-1	8-9	9-4	9-11	10-6	11-0	11-6	11-11	12-5	12-10	13-3	13-8	14-0	14-5	14-10	15-2	15-6	15-10	16-3
2x10	12.0	10-4	11-11	13-4	14-8	15-10	16-11	17-11	18-11	19-10	20-8	21-6	22-4	23-1	23-11	24-7	25-4	26-0					
	16.0	8-11	10-4	11-7	12-8	13-8	14-8	15-6	16-4	17-2	17-11	18-8	19-4	20-0	20-8	21-4	21-11	22-6	23-1	23-8	24-3		
	19.2	8-2	9-5	10-7	11-7	12-6	13-4	14-2	14-11	15-8	16-4	17-0	17-8	18-3	18-11	19-6	20-0	20-7	21-1	21-8	22-2	22-8	
	24.0	7-4	8-5	9-5	10-4	11-2	11-11	12-8	13-4	14-0	14-8	15-3	15-10	16-4	16-11	17-5	17-11	18-5	18-11	19-4	19-10	20-3	20-8
2x12	12.0	12-7	14-6	16-3	17-9	19-3	20-6	21-9	23-0	24-1	25-2												
	16.0	10-11	12-7	14-1	15-5	16-8	17-9	18-10	19-11	20-10	21-9	22-8	23-6	24-4	25-2								
	19.2	10-0	11-6	12-10	14-1	15-2	16-3	17-3	18-2	19-0	19-11	20-8	21-6	22-3	23-0	23-8	24-4	25-0	25-8				
	24.0	8-11	10-3	11-6	12-7	13-7	14-8	15-5	16-3	17-0	17-9	18-6	19-3	19-11	20-6	21-2	21-9	22-5	23-0	23-6	24-1	24-8	25-2
E	12.0	0.15	0.23	0.32	0.43	0.54	0.66	0.78	0.92	1.06	1.21	1.36	1.52	1.69	1.86	2.04	2.22	2.41	2.60				
	16.0	0.13	0.20	0.28	0.37	0.47	0.57	0.68	0.80	0.92	1.05	1.18	1.32	1.46	1.61	1.76	1.92	2.08	2.25	2.42	2.60		
	19.2	0.12	0.18	0.26	0.34	0.43	0.52	0.62	0.73	0.84	0.95	1.08	1.20	1.33	1.47	1.61	1.75	1.90	2.05	2.21	2.37	2.53	
	24.0	0.11	0.18	0.23	0.30	0.38	0.46	0.55	0.65	0.75	0.85	0.96	1.08	1.19	1.31	1.44	1.57	1.70	1.84	1.98	2.12	2.27	2.41

NOTE: The required modulus of elasticity, E, in 1,000,000 pounds per square inch (psi) (× 0.00689 for N/mm²) is shown at the bottom of this table, is limited to 2.6 million psi (17 914 N/mm²) and less, and is applicable to all lumber sizes shown. Spans are shown in feet-inches (1 foot = 304.8 mm, 1 inch = 25.4 mm) and are limited to 26 feet (7925 mm) and less.

TABLE B.13 Allowable Spans in Feet for Rafters—Live Load 20 psf, Dead Load 20 psf, Live Load Deflection L/240 (UBC Table 23-I-V-R-5)

DESIGN CRITERIA:
Strength — Live load of 20 psf (0.96 kN/m²) plus dead load of 20 psf (0.96 kN/m²) determines the required bending design value.
Deflection — For 20 psf (0.96 kN/m²) live load.
Limited to span in inches (mm) divided by 240.

Rafter Size (in) × 25.4 for mm	Spacing (in)	Bending Design Value, F_b (psi) × 0.00689 for N/mm²																								
		300	400	500	600	700	800	900	1000	1100	1200	1300	1400	1500	1600	1700	1800	1900	2000	2100	2200	2300	2400	2500	2600	2700
2 x 6	12.0	8-2	7-1	7-11	8-8	9-5	10-0	10-8	11-3	11-9	12-4	12-10	13-3	13-8	14-2	14-8	15-1	15-6	15-11	16-3	16-8	17-0	17-5	17-8	18-1	18-1
	16.0	5-4	6-2	6-10	7-6	8-2	8-8	9-3	9-8	10-2	10-8	11-1	11-6	11-11	12-4	12-8	13-1	13-5	13-9	14-1	14-5	14-8	15-1	15-4	15-8	16-0
	19.2	4-10	5-7	6-3	6-10	7-5	7-11	8-5	8-11	9-4	9-8	10-1	10-6	10-10	11-3	11-7	11-11	12-3	12-7	12-10	13-2	13-6	13-9	14-0	14-4	14-7
	24.0	4-4	5-0	5-7	6-2	6-8	7-1	7-6	7-11	8-4	8-8	9-1	9-5	9-8	10-0	10-4	10-8	10-11	11-3	11-6	11-9	12-0	12-4	12-7	12-10	13-1
2 x 8	12.0	8-1	8-4	10-6	11-6	12-5	13-3	14-0	14-10	15-6	16-3	16-10	17-6	18-1	18-8	19-4	19-10	20-5	20-11	21-5	21-11	22-5	22-11	23-5	23-8	24-1
	16.0	7-0	8-1	9-1	9-11	10-9	11-6	12-2	12-10	13-5	14-0	14-7	15-2	15-8	16-2	16-9	17-2	17-8	18-1	18-7	19-0	19-5	19-10	20-3	20-8	21-1
	19.2	6-5	7-5	8-3	9-1	9-10	10-6	11-1	11-8	12-3	12-10	13-4	13-10	14-4	14-10	15-3	15-8	16-2	16-7	16-11	17-4	17-9	18-0	18-1	18-6	21-0
	24.0	5-9	6-7	7-5	8-1	8-9	9-4	9-11	10-5	11-0	11-6	11-11	12-5	12-10	13-3	13-8	14-0	14-5	14-10	15-2	15-6	15-10	16-3	16-7	16-10	17-2
2 x 10	12.0	10-4	11-11	13-4	14-8	15-10	16-11	17-11	18-11	19-10	20-8	21-6	22-4	23-1	23-11	24-7	25-4	26-0	23-1	23-8	24-3	24-10	25-4	25-10	24-1	24-8
	16.0	8-11	10-4	11-7	12-8	13-8	14-8	15-6	16-4	17-2	17-11	18-8	19-4	20-0	20-8	21-4	21-11	22-6	23-1	23-8	24-3	24-10	23-1	23-7	24-0	21-1
	19.2	8-2	9-5	10-7	11-7	12-6	13-4	14-2	14-11	15-8	16-4	17-0	17-8	18-3	18-11	19-6	20-0	20-7	21-1	21-8	22-2	22-8	23-1	23-1	23-8	20-3
	24.0	7-4	8-5	9-5	10-4	11-2	11-11	12-8	13-4	14-0	14-8	15-3	15-10	16-4	16-11	17-5	17-11	18-5	18-11	19-4	19-10	19-10	20-3	21-1	21-6	21-11
2 x 12	12.0	12-7	14-6	16-3	17-8	19-3	20-6	21-9	23-0	24-1	25-2	22-8	23-8	23-1	23-11	25-11	25-4	26-0		23-8	24-3	24-10	25-4	25-10		
	16.0	10-11	12-7	14-1	15-5	16-8	17-9	18-10	19-11	20-10	21-9	22-8	23-8	24-4	25-2	25-11	24-4	25-0	25-8							
	19.2	9-11	11-6	12-10	14-1	15-2	16-3	17-3	18-2	19-0	19-11	20-8	21-5	22-3	23-0	23-8	21-8	22-7	23-6	21-6	22-2	22-8		25-2		
	24.0	8-11	10-3	11-6	12-7	13-7	14-6	15-5	16-3	17-0	17-9	18-6	19-3	19-11	20-8	21-2	21-8	22-5	22-11	20-4	24-1	24-8	24-0	25-2		
E	12.0	0.10	0.15	0.22	0.29	0.36	0.44	0.52	0.61	0.71	0.80	0.91	1.01	1.13	1.24	1.36	1.48	1.60	1.73	1.86	2.00	2.14	2.28	2.42	2.57	2.35
	16.0	0.09	0.13	0.19	0.25	0.31	0.38	0.45	0.53	0.61	0.70	0.79	0.88	0.97	1.07	1.18	1.29	1.39	1.50	1.61	1.73	1.85	1.97	2.10	2.22	2.15
	19.2	0.08	0.12	0.17	0.23	0.29	0.35	0.41	0.48	0.56	0.64	0.72	0.80	0.89	0.98	1.07	1.17	1.27	1.37	1.47	1.58	1.69	1.80	1.91	2.03	2.08
	24.0	0.07	0.11	0.15	0.20	0.25	0.31	0.37	0.43	0.50	0.57	0.64	0.72	0.80	0.88	0.96	1.05	1.13	1.22	1.32	1.41	1.51	1.61	1.71	1.82	1.92

NOTE: The required modulus of elasticity, E, in 1,000,000 pounds per square inch (psi) (× 0.00689 for N/mm²) is shown at the bottom of this table, is limited to 2.6 million psi (17 914 N/mm²) and less, and is applicable to all lumber sizes shown. Spans are shown in feet-inches (1 foot = 304.8 mm, 1 inch = 25.4 mm) and are limited to 26 feet (7925 mm) and less.

TABLE B.14 Allowable Spans in Feet for Rafters—Live Load 30 psf, Dead Load 20 psf, Live Load Deflection L/240 (UBC Table 23-I-V-R-6)

DESIGN CRITERIA:
Strength — Live load of 30 psf (1.44 kN/m²) plus dead load of 20 psf (0.96 kN/m²) determines the required bending design value.
Deflection — For 30 psf (1.44 kN/m²) live load.
Limited to span in inches (mm) divided by 240.

Rafter Size (in)	Spacing (in) × 25.4 for mm	|| Bending Design Value, F_b (psi) × 0.00689 for N/mm² ||																							
		300	400	500	600	700	800	900	1000	1100	1200	1300	1400	1500	1600	1700	1800	1900	2000	2100	2200	2300	2400	2500	2600	2700
2 x 6	12.0	5-8	6-4	7-1	7-9	8-5	8-11	8-8	10-0	10-6	11-0	11-5	11-11	12-4	12-8	13-1	13-6	13-10	14-2	14-7	14-11	15-3	15-7	15-11	14-8	14-3
	16.0	4-11	5-8	6-2	6-9	7-3	7-9	8-2	8-8	9-1	9-6	9-11	10-3	10-8	11-0	11-4	11-8	12-0	12-4	12-7	12-11	13-2	13-6	13-9	14-0	14-3
	19.2	4-4	5-0	5-7	6-2	6-8	7-1	7-6	7-11	8-4	8-8	9-1	9-5	9-9	10-0	10-4	10-8	10-11	11-3	11-6	11-9	12-0	12-3	12-7	12-10	13-1
	24.0	3-11	4-6	5-0	5-6	5-11	6-4	6-8	7-1	7-5	7-9	8-1	8-5	8-8	9-0	9-3	9-6	9-9	10-0	10-3	10-6	10-8	10-11	11-2	11-5	11-7
2 x 8	12.0	7-3	8-4	9-4	10-3	11-1	11-10	12-7	13-3	13-11	14-6	15-1	15-8	16-2	16-9	17-3	17-9	18-3	18-8	19-2	19-8	20-1	20-6	20-11	18-8	18-10
	16.0	6-3	7-3	8-1	8-11	9-7	10-3	10-10	11-6	12-0	12-7	13-1	13-7	14-0	14-6	14-11	15-5	15-10	16-3	16-7	17-0	17-5	17-9	18-1	18-6	18-10
	19.2	5-9	6-7	7-5	8-1	8-9	9-4	9-11	10-6	11-0	11-6	11-11	12-5	12-10	13-3	13-8	14-0	14-5	14-10	15-2	15-6	15-10	16-2	16-7	16-10	17-2
	24.0	5-2	5-11	6-7	7-3	7-10	8-4	8-11	9-4	9-10	10-3	10-8	11-1	11-6	11-10	12-2	12-7	12-11	13-3	13-7	13-11	14-2	14-6	14-10	15-1	15-5
2 x 10	12.0	11-3	10-8	11-11	13-1	14-2	15-1	16-0	16-11	17-9	18-6	19-3	20-0	20-9	21-4	22-0	22-8	23-3	23-11	24-6	25-1	25-7	...	22-8	23-7	24-0
	16.0	8-0	9-3	10-4	11-4	12-3	13-1	13-10	14-8	15-4	16-0	16-8	17-4	17-11	18-6	19-1	19-7	20-2	20-8	21-2	21-8	22-2	22-8	23-1	23-7	24-0
	19.2	7-4	8-5	9-5	10-4	11-2	11-11	12-8	13-4	14-0	14-8	15-3	15-10	16-4	16-11	17-5	17-11	18-5	18-11	19-4	19-10	20-3	20-8	21-1	21-6	21-11
	24.0	6-6	7-7	8-5	9-3	10-0	10-8	11-4	11-11	12-6	13-1	13-7	14-2	14-8	15-1	15-7	16-0	16-6	16-11	17-4	17-8	18-1	18-6	18-11	19-3	19-7
2 x 12	12.0	11-3	13-0	14-6	15-11	17-2	18-4	18-10	20-6	21-7	22-6	23-5	24-4	25-2	26-0	26-10	27-8	28-6	29-2	...	29-11	...	...	25-2	25-8	...	29-10
	16.0	9-9	11-3	12-7	13-9	14-11	15-11	16-10	17-10	18-8	19-6	20-3	21-1	21-9	22-6	23-2	23-10	24-6	25-2	25-9	26-4	27-0	27-8	28-4	29-0	29-7
	19.2	8-11	10-3	11-6	12-7	13-7	14-6	15-5	16-3	17-0	17-10	18-6	19-3	19-11	20-6	21-2	21-9	22-5	23-0	23-6	24-1	24-8	25-2	25-8	26-2	26-8
	24.0	7-11	9-2	10-3	11-3	12-2	13-0	13-9	14-6	15-3	15-11	16-7	17-2	17-9	18-4	18-11	19-6	20-0	20-6	21-1	21-7	22-0	22-6	23-0	23-5	23-10
E	12.0	0.11	0.17	0.22	0.31	0.38	0.47	0.56	0.66	0.76	0.86	0.97	1.08	1.21	1.33	1.46	1.59	1.72	1.86	2.00	2.14	2.29	2.44	2.59		2.53
	16.0	0.09	0.14	0.20	0.26	0.33	0.41	0.48	0.57	0.66	0.75	0.84	0.94	1.05	1.15	1.26	1.37	1.49	1.61	1.73	1.86	1.98	2.12	2.25	2.39	2.53
	19.2	0.08	0.13	0.18	0.24	0.30	0.37	0.44	0.52	0.60	0.68	0.77	0.86	0.95	1.05	1.15	1.25	1.36	1.47	1.58	1.70	1.81	1.93	2.05	2.18	2.31
	24.0	0.06	0.12	0.16	0.22	0.27	0.33	0.40	0.46	0.54	0.61	0.68	0.77	0.85	0.94	1.03	1.12	1.22	1.31	1.41	1.52	1.62	1.73	1.84	1.95	2.06

NOTE: The required modulus of elasticity, E, in 1,000,000 pounds per square inch (psi) (× 0.00689 for N/mm²) is shown at the bottom of this table, is limited to 2.6 million psi (17 914 N/mm²) and less, and is applicable to all lumber sizes shown. Spans are shown in feet-inches (1 foot = 304.8 mm, 1 inch = 25.4 mm) and are limited to 26 feet (7925 mm) and less.

source: the 1994 edition of the UBC. For the reader's convenience, the original UBC table numbers are also given.

In using this data the reader should be careful to read all the qualifying conditions noted with the tables. True design conditions may often not fall exactly into the table qualifications. Nevertheless, the tables can be used for a quick approximate selection. In some cases it may be necessary to consult two tables that bracket the true conditions to see the range of selections between which a correct selection may lie.

Examples of the use of this data are given in various illustrations in the book.

B.3   STEEL STRUCTURES

Figure B.3 provides data relating to the deflection of rolled steel beams (A36 grade steel). However, the graphs are based on a maximum bending stress of 24 ksi, which is the limit but seldom the actual design situation. When determining the actual dimensions of deflection, you should keep the true stress condition in mind. Actually, the graphs are more useful for preliminary design, using the lines for common limits of span/deflection ratios to determine critical depth/span ratios.

Table B.15 can help you determine approximate capacities of ordinary formed sheet steel decks for roofs. When used for floors, these decks are covered with a cast concrete fill, which is typically considered to interact with the deck for both stress considerations and deflection. Data for floor decks should be obtained from deck manufacturers.

Table B.16 can help you determine approximate sizes for a common type of steel open-web joist: light, manufactured steel truss. Sometimes open-web joists are used in combination with supporting trusses, called *joist girders*. Table data is abstracted from literature of the Steel Joist Institute. The following discussion explains the use of table data and the process for design of both open-web joists and joist girders.

Parallel-chord trusses are produced in a wide range of sizes with various details by a number of manufacturers. Most producers comply with the regulations developed by the Steel Joist Institute (SJI), whose publications are a chief source of design information (for example, see Ref. 8). The products of individual manufacturers may vary, however, so information may be provided by suppliers or the manufacturers themselves.

The smallest and lightest products, called *open-web joists*, are used to directly support roof and floor decks, sustaining only uniformly distributed loads on their top chords. A popular form of open-web joist consists of chords of cold-formed sheet steel and webs of steel rods. Chords may also be double angles, with the rods sandwiched between the angles at joints.

Table B.16, adapted from standard SJI tables, lists the range of joist sizes available in the basic K series. (*Note:* A few of the heavier sizes have been omitted.) Joists are identified by a three-unit designation: the first number

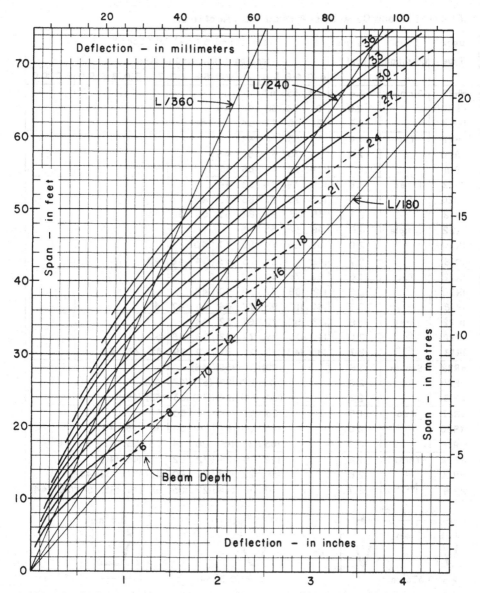

FIGURE B.3 Deflection of steel beams with simple spans and uniformly distributed loading. Assumed conditions: A36 steel, maximum bending stress of 24 ksi.

indicates the overall depth of the joist, the letter tells the series, and the second number gives the class of size of the members used. The higher the number, the heavier the joist.

You can use Table B.16 to select the proper joist for a determined load and span condition. There are usually two entries in the table for each span: the first number represents the total load capacity of the joist, and the number in parentheses identifies the load that will produce a deflection of $\frac{1}{360}$ of the span.

TABLE B.15   Load Capacity of Formed Steel Roof Deck

Deck[b] Type	Span Condition	Weight[c] (psf)	Total (Dead & Live) Safe Load[d] for Spans Indicated in ft-in.												
			4-0	4-6	5-0	5-6	6-0	6-6	7-0	7-6	8-0	8-6	9-0	9-6	10-0
NR22	Simple	1.6	73	58	47										
NR20		2.0	91	72	58	48	40								
NR18		2.7	121	95	77	64	54	46							
NR22	Two	1.6	80	63	51	42									
NR20		2.0	96	76	61	51	43								
NR18		2.7	124	98	79	66	55	47	41						
NR22	Three or More	1.6	100	79	64	53	44								
NR20		2.0	120	95	77	63	53	45							
NR18		2.7	155	123	99	82	69	59	51	44					
IR22	Simple	1.6	86	68	55	45									
IR20		2.0	106	84	68	56	47	40							
IR18		2.7	142	112	91	75	63	54	46	40					
IR22	Two	1.6	93	74	60	49	41								
IR20		2.0	112	88	71	59	50	42							
IR18		2.7	145	115	93	77	64	55	47	41					
IR22	Three or More	1.6	117	92	75	62	52	44							
IR20		2.0	140	110	89	74	62	53	46	40					
IR18		2.7	181	143	116	96	81	69	59	52	45	40			
WR22	Simple	1.6			(89)	(70)	(56)	(46)							
WR20		2.0			(112)	(87)	(69)	(57)	(47)	(40)					
WR18		2.7			(154)	(119)	(94)	(76)	(63)	(53)	(45)				
WR22	Two	1.6			98	81	68	58	50	43					
WR20		2.0			125	103	87	74	64	55	49	43			
WR18		2.7			165	137	115	98	84	73	65	57	51	46	41
WR22	Three or More	1.6			122	101	85	72	62	54	(46)	(40)			
WR20		2.0			156	129	108	92	80	(67)	(57)	(49)	(43)		
WR18		2.7			207	171	144	122	105	(91)	(76)	(65)	(57)	(50)	(44)

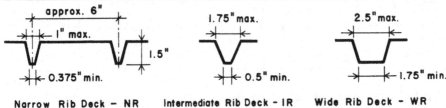

approx. 6"
1" max.
1.5"
0.375" min.

1.75" max.
0.5" min.

2.5" max.
1.75" min.

Narrow Rib Deck — NR   Intermediate Rib Deck - IR   Wide Rib Deck - WR

For heavier loads and longer spans, trusses are produced in two series, described as long span and deep long span, the latter achieving depths of 7 ft and spans approaching 150 ft. In some situations the loading and span may clearly indicate the choice of the series, as well as the specific size of member. In many cases, however, the series overlap, so the choice depends on the product costs.

Open-web, long-span, and deep long-span trusses are essentially designed for the uniformly loaded condition. This load may be due only to a

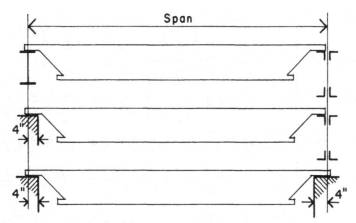

FIGURE B.4 Definition of span for open-web steel beams, as given in Table B.16.

roof or floor deck on the top, or it may include a ceiling attached to the bottom. For roofs, an often-used potential is that for sloping the top chord to facilitate drainage, while maintaining the flat bottom for a ceiling. Relatively small concentrated loads may be tolerated, especially if they are applied close to the truss joints.

Load tables for all standard products, as well as specifications for lateral bracing, support details, and so on, are available from the SJI. Planar trusses, like very thin beams, need considerable lateral support. Attached roof or floor decks and even ceilings may provide the necessary support if the attachment and the stiffness of the bracing construction are adequate. For trusses where no ceiling exists, you must use bridging or other forms of lateral bracing.

Bracing is a three-dimensional problem, involving not only the stability of the trusses, but the general development of the building construction. Elements used for bracing can often serve as supports for ducts, lighting, building equipment, catwalks, and so on.

To develop a complete truss system, you can use a special type of prefabricated truss, called a *joist girder*. This truss is specifically designed to carry the regularly spaced, concentrated loads consisting of the end supports of joists.

Figure B.5 shows the general joist girder form and the joist girder standard designation, which includes indications of the nominal girder depth, weight, spacing of members, and the unit load to be carried.

You can select predesigned joist girders from manufacturer catalogs just as you do open-web joists. The procedure is usually as follows:

1. Determine the desired joist spacing loads to be carried, and girder span. The joist spacing must be an even-number division of the girder span.

TABLE B.16 Allowable Loads for K-Series Open Web Steel Joists[a]

Joist Designation	12K1	12K3	12K5	14K1	14K3	14K4	14K6	16K2	16K3	16K4	16K6	18K3	18K4	18K5	18K7	20K3	20K4	20K5	20K7
Weight (lb/ft)	5.0	5.7	7.1	5.2	6.0	6.7	7.7	5.5	6.3	7.0	8.1	6.6	7.2	7.7	9.0	6.7	7.6	8.2	9.3
Span (ft)																			
20	241 (142)	302 (177)	409 (230)	284 (197)	356 (246)	428 (287)	525 (347)	368 (297)	410 (330)	493 (386)	550 (426)	463 (423)	550 (490)	550 (490)	550 (490)	517 (517)	550 (550)	550 (550)	550 (550)
22	199 (106)	249 (132)	337 (172)	234 (147)	293 (184)	353 (215)	432 (259)	303 (222)	337 (247)	406 (289)	498 (351)	382 (316)	460 (370)	518 (414)	550 (438)	426 (393)	514 (461)	550 (490)	550 (490)
24	166 (81)	208 (101)	282 (132)	196 (113)	245 (141)	295 (165)	362 (199)	254 (170)	283 (189)	340 (221)	418 (269)	320 (242)	385 (284)	434 (318)	526 (382)	357 (302)	430 (353)	485 (396)	550 (448)
26				166 (88)	209 (110)	251 (129)	308 (156)	216 (133)	240 (148)	289 (173)	355 (211)	272 (190)	328 (222)	369 (249)	448 (299)	304 (236)	366 (277)	412 (310)	500 (373)
28				143 (70)	180 (88)	216 (103)	265 (124)	186 (106)	207 (118)	249 (138)	306 (168)	234 (151)	282 (177)	318 (199)	385 (239)	261 (189)	315 (221)	355 (248)	430 (298)
30								161 (86)	180 (96)	216 (112)	266 (137)	203 (123)	245 (144)	276 (161)	335 (194)	227 (153)	274 (179)	308 (201)	374 (242)
32								142 (71)	158 (79)	190 (92)	233 (112)	178 (101)	215 (118)	242 (132)	294 (159)	199 (126)	240 (147)	271 (165)	328 (199)
36												141 (70)	169 (82)	191 (92)	232 (111)	157 (88)	189 (103)	213 (115)	259 (139)
40																127 (64)	153 (75)	172 (84)	209 (101)

Joist Designation	22K4	22K5	22K6	22K9	24K4	24K5	24K6	24K9	26K5	26K6	26K9	28K6	28K7	28K8	28K10	30K7	30K8	30K9	30K12
Weight (lb/ft)	8.0	8.8	9.2	11.3	8.4	9.3	9.7	12.0	9.8	10.6	12.2	11.4	11.8	12.7	14.3	12.3	13.2	13.4	17.6
Span (ft)																			
28	348 (270)	392 (302)	427 (328)	550 (413)	381 (323)	429 (362)	467 (393)	550 (456)	466 (427)	508 (464)	550 (501)	548 (541)	550 (543)	550 (543)	550 (543)				
30	302 (219)	341 (245)	371 (266)	497 (349)	331 (262)	373 (293)	406 (319)	544 (419)	405 (346)	441 (377)	550 (459)	477 (439)	531 (486)	550 (500)	550 (500)	550 (543)	550 (543)	550 (543)	550 (543)
32	265 (180)	299 (201)	326 (219)	436 (287)	290 (215)	327 (241)	357 (262)	478 (344)	356 (285)	387 (309)	519 (407)	418 (361)	466 (400)	515 (438)	549 (463)	501 (461)	549 (500)	549 (500)	549 (500)
36	209 (126)	236 (141)	257 (153)	344 (201)	229 (150)	258 (169)	281 (183)	377 (241)	280 (199)	305 (216)	409 (284)	330 (252)	367 (280)	406 (306)	487 (366)	395 (323)	436 (353)	475 (383)	487 (392)
40	169 (91)	190 (102)	207 (111)	278 (146)	185 (109)	208 (122)	227 (133)	304 (175)	227 (145)	247 (157)	331 (207)	266 (183)	297 (203)	328 (222)	424 (284)	319 (234)	353 (256)	384 (278)	438 (315)
44	139 (68)	157 (76)	171 (83)	229 (109)	153 (82)	172 (92)	187 (100)	251 (131)	187 (108)	204 (118)	273 (155)	220 (137)	245 (152)	271 (167)	350 (212)	263 (176)	291 (192)	317 (208)	398 (258)
48					128 (63)	144 (70)	157 (77)	211 (101)	157 (83)	171 (90)	229 (119)	184 (105)	206 (117)	227 (128)	294 (163)	221 (135)	244 (148)	266 (160)	365 (216)
52									133 (65)	145 (71)	195 (93)	157 (83)	175 (92)	193 (100)	250 (128)	188 (106)	208 (116)	226 (126)	336 (184)
56												135 (66)	151 (73)	166 (80)	215 (102)	162 (84)	179 (92)	195 (100)	301 (153)
60																141 (69)	156 (75)	169 (81)	262 (124)

[a] Loads in pounds per foot of joist span; first entry represents the total joist capacity, entry in parentheses is the load that produces a maximum deflection of $\frac{L}{360}$ of the span. See Figure B.4 for definition of span.

Source: Data adapted from more extensive tables in the *Standard Specifications, Load Tables, and Weight Tables for Steel Joists and Joist Girders*, 1986 ed. (Ref. 8), with permission of the publisher, Steel Joist Institute. The Steel Joist Institute publishes both specifications and load tables; each contains standards that are to be used in conjunction with one another.

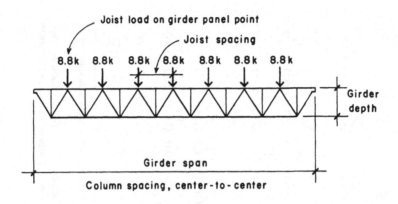

FIGURE B.5 Method of standard designation for steel joist girders.

2. Determine the total design load for one supported joist. This is the unit concentrated load on the girder.
3. Find the appropriate girder weight and depth from design tables in a manufacturer's catalog.

Table B.17 can help you determine capacities of steel beams consisting of rolled steel shapes (mostly W shapes). The table values assume a maximum allowable bending stress of 24 ksi.

Table B.18 provides values for the axial compression capacity of steel columns consisting of the most widely used W shapes of A36 steel. When choosing steel columns, you need to accommodate the framing of steel beams into the column in both the direction that faces the column flanges and the direction that faces the column web. You generally need a flange width of at least 6 in. and a nominal depth for shapes of at least 10 in.

TABLE B.17 Load/Span Values for Rolled Steel Beams[a]

Shape	Span (ft) → L_c** (ft) / Deflection Factor*	8	10	12	14	16	18	20	22	24	26	28	30
		1.59	2.48	3.58	4.87	6.36	8.05	9.93	12.0	14.3	16.8	19.5	22.3
M 8 × 6.5	2.4	9.24	7.39	6.16	5.28	4.62	4.11						
M 10 × 9	2.6	15.5	12.4	10.3	8.87	7.76	6.90	6.21	5.64				
W 8 × 10	4.2	15.6	12.5	10.4	8.92	7.81	6.94						
W 8 × 13	4.2	19.8	15.9	13.2	11.3	9.91	8.81						
W 10 × 12	3.9	21.8	17.4	14.5	12.5	10.9	9.69	8.72	7.93				
W 8 × 15	4.2	23.6	18.9	15.7	13.5	11.8	10.5						
M 12 × 11.8	2.7	24.0	19.2	16.0	13.7	12.0	10.7	9.60	8.73	8.00	7.38	6.86	
W 10 × 15	4.2	27.6	22.1	18.4	15.8	13.8	12.3	11.0	10.0				
W 12 × 14	3.5	29.8	23.8	19.9	17.0	14.9	13.2	11.9	10.8	9.93	9.17	8.51	
W 8 × 18	5.5	30.4	24.3	20.3	17.4	15.2	13.5						
W 10 × 17	4.2	32.4	25.9	21.6	18.5	16.2	14.4	13.0	11.8				
W 12 × 16	4.1	34.2	27.4	22.8	19.5	17.1	15.2	13.7	12.4	11.4	10.5	9.77	
W 8 × 21	5.6	36.4	29.1	24.3	20.8	18.2	16.2						
W 10 × 19	4.2	37.6	30.1	25.1	21.5	18.8	16.7	15.0	13.7				
W 8 × 24	6.9	41.8	33.4	27.9	23.9	20.9	18.6						
M 14 × 18	3.6	42.2	33.8	28.1	24.1	21.1	18.7	16.9	15.3	14.1	13.0	12.0	11.2
W 12 × 19	4.2	42.6	34.1	28.4	24.3	21.3	18.9	17.0	15.5	14.2	13.1	12.2	
W 10 × 22	6.1	46.4	37.1	30.9	26.5	23.2	20.6	18.5	16.9				
W 8 × 28	6.9	48.6	38.9	32.4	27.8	24.3	21.6						

Shape	Span (ft) → L_c** (ft) / Deflection Factor*	12	14	16	18	20	22	24	26	28	30	32	34
		3.58	4.87	6.36	8.05	9.93	12.0	14.3	16.8	19.5	22.3	25.4	28.7
W 12 × 22	4.3	33.9	29.0	25.4	22.6	20.3	18.5	16.9	15.6	14.5			
W 10 × 26	6.1	37.2	31.9	27.9	24.8	22.3	20.3						
W 14 × 22	5.3	38.7	33.1	29.0	25.8	23.2	21.1	19.3	17.8	16.6	15.5	14.5	
W 10 × 30	6.1	43.2	37.0	32.4	28.8	25.9	23.6						
W 12 × 26	6.9	44.5	38.2	33.4	29.7	26.7	24.3	22.3	20.5	19.1			
W 10 × 33	8.4	46.7	40.0	35.0	31.0	28.0	25.4						
W 14 × 26	5.3	47.1	40.3	35.3	31.4	28.2	25.7	23.5	21.7	20.2	18.8	17.6	
W 16 × 26	5.6	51.2	43.9	38.4	34.1	30.7	27.9	25.6	23.6	21.9	20.5	19.2	18.1
W 12 × 30	6.9	51.5	44.1	38.6	34.3	30.9	28.1	25.7	23.8	22.0			
W 14 × 30	7.1	56.0	48.0	42.0	37.3	33.6	30.5	28.0	25.8	24.0	22.4	21.0	
W 10 × 39	8.4	56.1	48.1	42.1	37.4	33.7	30.6						
W 12 × 35	6.9	60.8	52.1	45.6	40.5	36.5	33.2	30.4	28.1	26.0			
W 16 × 31	5.8	62.9	53.9	47.2	41.9	37.8	34.3	31.5	29.0	27.0	25.2	23.6	22.2
W 14 × 34	7.1	64.8	55.5	48.6	43.2	38.9	35.3	32.4	29.9	27.8	25.9	24.3	
W 10 × 45	8.5	65.5	56.1	49.1	43.6	39.3	35.7						

Shape	Span (ft) → L_c** (ft) / Deflection Factor*	16	18	20	22	24	26	28	30	32	34	36	38
		6.36	8.05	9.93	12.0	14.3	16.8	19.5	22.3	25.4	28.7	32.2	35.9
W 12 × 40	8.4	51.9	46.1	41.5	37.7	34.6	31.9	29.6					
W 14 × 38	7.1	54.6	48.5	43.7	39.7	36.4	33.6	31.2	29.1	27.3			
W 16 × 36	7.4	56.5	50.2	45.2	41.1	37.7	34.8	32.3	30.1	28.2	26.6	25.1	
W 18 × 35	6.3	57.8	51.4	46.2	42.0	38.5	35.6	33.0	30.8	28.9	27.2	25.7	24.3
W 12 × 45	8.5	58.1	51.6	46.5	42.2	38.7	35.7	33.2					
W 14 × 43	8.4	62.7	55.7	50.1	45.6	41.8	38.6	35.8	33.4	31.3			
W 12 × 50	8.5	64.7	57.5	51.7	47.0	43.1	39.8	37.0					
W 16 × 40	7.4	64.7	57.5	51.7	47.0	43.1	39.8	37.0	34.5	32.3	30.4	28.7	
W 18 × 40	6.3	68.4	60.8	54.7	49.7	45.6	42.1	39.1	36.5	34.2	32.2	30.4	28.8
W 14 × 48	8.5	70.3	62.5	56.2	51.1	46.9	43.3	40.2	37.5	35.1			
W 12 × 53	10.6	70.6	62.7	56.5	51.3	47.1	43.4	40.3					
W 16 × 45	7.4	72.7	64.6	58.2	52.9	48.5	44.7	41.5	38.8	36.3	34.2	32.3	
W 14 × 53	8.5	77.8	69.1	62.2	56.6	51.9	47.9	44.4	41.5	38.9			
W 18 × 46	6.4	78.8	70.0	63.0	57.3	52.5	48.5	45.0	42.0	39.4	37.1	35.0	33.2
W 16 × 50	7.5	81.0	72.0	64.8	58.9	54.0	49.8	46.3	43.2	40.5	38.1	36.0	

[a] Total allowable uniformly distributed load is in kips for simple span beams of A36 steel with yield stress of 36 ksi [250 MPa].

[b] Maximum deflection in inches at the center of the span may be obtained by dividing this factor by the depth of the beam in inches. This is based on a maximum bending stress of 24 ksi [165 MPa].

[c] Maximum permitted distance between points of lateral support. If distance exceeds this, use the charts in the AISC Manual.

(continued)

TABLE B.17 (Continued)

Shape	L_c** (ft)	Span (ft) 16	18	20	22	24	27	30	33	36	39	42	45
	Deflection Factor*	6.36	8.05	9.93	12.0	14.3	18.1	22.3	27.0	32.2	37.8	43.8	50.3
W 21 × 44	6.6	81.6	72.5	65.3	59.3	54.4	48.3	43.5	39.6	36.3	33.5	31.1	29.0
W 18 × 50	7.9	88.9	79.0	71.1	64.6	59.3	52.7	47.4	43.1	39.5	36.5		
W 14 × 61	10.6	92.2	81.9	73.8	67.0	61.5	54.6	49.2	44.7				
W 16 × 57	7.5	92.2	81.9	73.8	67.0	61.5	54.6	49.2	44.7	41.0			
W 21 × 50	6.9	94.5	84.0	75.6	68.7	63.0	56.0	50.4	45.8	42.0	38.8	36.0	33.6
W 18 × 55	7.9	98.3	87.4	78.6	71.5	65.5	58.2	52.4	47.7	43.7	40.3		
W 18 × 60	8.0	108	96.0	86.4	78.5	72.0	64.0	57.6	52.4	48.0	44.3		
W 21 × 57	6.9	111	98.7	88.6	80.7	74.0	65.8	59.2	53.8	49.3	45.5	42.3	39.5
W 24 × 55	7.0	114	101	91.2	82.9	76.0	67.5	60.8	55.3	50.7	46.8	43.4	40.5
W 16 × 67	10.8	117	104	93.6	85.1	78.0	69.3	62.4	56.7	52.0			
W 18 × 65	8.0	117	104	93.6	85.1	78.0	69.3	62.4	56.7	52.0	48.0		
W 18 × 71	8.1	127	113	102	92.4	84.7	72.2	67.7	61.5	56.4	52.1		
W 21 × 62	8.7	127	113	102	92.4	84.7	72.2	67.7	61.5	56.4	52.1	48.4	45.1
W 24 × 62	7.4	131	116	105	95.3	87.3	77.6	69.9	63.5	58.2	53.7	49.9	46.6
W 16 × 77	10.9	134	119	107	97.4	89.3	79.4	71.5	65.0	59.5			
W 21 × 68	8.7	140	124	112	102	93.3	83.0	74.7	67.9	62.2	57.4	53.3	49.8
W 18 × 76	11.6	146	130	117	106	97.3	86.5	77.9	70.8	64.9	59.9		
W 21 × 73	8.8	151	134	121	110	101	89.5	80.5	73.2	67.1	61.9	57.5	53.7
W 24 × 68	9.5	154	137	123	112	103	91.2	82.1	74.7	68.4	63.2	58.7	54.7
W 18 × 86	11.7	166	147	133	121	111	98.4	88.5	80.5	73.8	68.1		
W 21 × 83	8.8	171	152	137	124	114	101	91.2	82.9	76.0	70.1	65.1	60.8

Shape	L_c** (ft)	Span (ft) 24	27	30	33	36	39	42	45	48	52	56	60
	Deflection Factor*	14.3	18.1	22.3	27.0	32.2	37.8	43.8	50.3	57.2	67.1	77.9	89.4
W 24 × 76	9.5	117	104	93.9	85.3	78.2	72.2	67.0	62.6	58.7			
W 21 × 93	8.9	128	114	102	93.1	85.3	78.8	73.1	68.3				
W 24 × 84	9.5	131	116	104	95.0	87.1	80.4	74.7	69.7	65.3			
W 27 × 84	10.5	142	126	114	103	94.7	87.4	81.1	75.7	71.0	65.5	60.8	
W 24 × 94	9.6	148	131	118	108	98.7	91.1	84.6	78.9	74.0			
W 21 × 101	13.0	151	134	121	110	101	93.1	86.5	80.7				
W 27 × 94	10.5	162	144	130	118	108	99.7	92.6	86.4	81.0	74.8	69.4	
W 24 × 104	13.5	172	153	138	125	115	106	98.3	91.7	86.0			
W 27 ×·102	10.6	178	158	142	129	119	109	102	94.9	89.0	82.1	76.3	
W 30 × 99	10.9	179	159	143	130	120	110	102	95.6	89.7	82.8	76.9	71.7
W 24 × 117	13.5	194	172	155	141	129	119	111	103	97.0			
W 27 × 114	10.6	199	177	159	145	133	123	114	106	99.7	92.0	85.4	
W 30 × 108	11.1	199	177	159	145	133	123	114	106	99.7	92.0	85.4	79.7
W 30 × 116	11.1	219	195	175	159	146	135	125	117	110	101	94.0	87.7
W 30 × 124	11.1	237	210	189	172	158	146	135	126	118	109	101	94.7

Shape	L_c** (ft)	Span (ft) 30	33	36	39	42	45	48	52	56	60	65	70
	Deflection Factor*	22.3	27.0	32.2	37.8	43.8	50.3	57.2	67.1	77.9	89.4	105	122
W 33 × 118	12.0	191	174	159	147	137	128	120	110	103	95.7	88.4	
W 30 × 132	11.1	203	184	169	156	145	135	127	117	109	101		
W 33 × 130	12.1	216	197	180	166	155	144	135	125	116	108	99.9	
W 27 × 146	14.7	219	199	183	169	156	146	137	126	117			
W 36 × 135	12.3	234	213	195	180	167	156	146	135	125	117	108	100
W 33 × 141	12.2	239	217	199	184	171	159	149	138	128	119	110	
W 33 × 152	12.2	260	236	216	200	185	173	162	150	139	130	120	
W 36 × 150	12.6	269	244	224	207	192	179	168	155	144	134	124	115
W 30 × 173	15.8	287	261	239	221	205	192	180	166	154	144		
W 36 × 160	12.7	289	263	241	222	206	193	181	167	155	144	133	124
W 36 × 170	12.7	309	281	258	238	221	206	193	178	166	155	143	132
W 30 × 191	15.9	319	290	268	245	228	213	199	184	171	159		
W 36 × 182	12.7	332	302	277	256	237	221	208	192	178	166	153	142
W 36 × 194	12.8	354	322	295	272	253	236	221	204	190	177	163	152
W 33 × 201	16.6	365	332	304	281	260	243	228	210	195	182	168	
W 36 × 210	12.9	383	349	319	295	274	256	240	221	205	192	177	164
W 33 × 221	16.7	404	367	336	310	288	269	252	233	216	202	186	
W 33 × 241	16.7	442	402	368	340	316	295	276	255	237	221	204	
W 36 × 230	17.4	446	406	372	343	319	298	279	257	239	223	206	191
W 36 × 245	17.4	477	434	398	367	341	318	298	275	256	239	220	204
W 36 × 260	17.5	508	462	423	391	363	339	318	293	272	254	234	218
W 36 × 280	17.5	549	499	458	422	392	366	343	317	294	275	253	235
W 36 × 300	17.6	592	538	493	455	423	395	370	341	317	296	273	254

TABLE B.18 Axial Compression Load Capacity of W-Shape Columns of A36 Steel

Shape	Effective length (KL) in feet										Bending factor	
	8	9	10	11	12	14	16	18	20	22	B_x	B_y
M 4 × 13	48	42	35	29	24	18					0.727	2.228
W 4 × 13	52	46	39	33	28	20	16				0.701	2.016
W 5 × 16	74	69	64	58	52	40	31	24	20		0.550	1.560
M 5 × 18.9	85	78	71	64	56	42	32	25			0.576	1.768
W 5 × 19	88	82	76	70	63	48	37	29	24		0.543	1.526
W 6 × 9	33	28	23	19	16	12					0.482	2.414
W 6 × 12	44	38	31	26	22	16					0.486	2.367
W 6 × 16	62	54	46	38	32	23	18				0.465	2.155
W 6 × 15	75	71	67	62	58	48	38	30	24	20	0.456	1.424
M 6 × 20	98	92	87	81	74	61	47	37	30	25	0.453	1.510
W 6 × 20	100	95	90	85	79	67	54	42	34	28	0.438	1.331
W 6 × 25	126	120	114	107	100	85	69	54	44	36	0.440	1.308
W 8 × 24	124	118	113	107	101	88	74	59	48	39	0.339	1.258
W 8 × 28	144	138	132	125	118	103	87	69	56	46	0.340	1.244
W 8 × 31	170	165	160	154	149	137	124	110	95	80	0.332	0.985
W 8 × 35	191	186	180	174	168	155	141	125	109	91	0.330	0.972
W 8 × 40	218	212	205	199	192	127	160	143	124	104	0.330	0.959
W 8 × 48	263	256	249	241	233	215	196	176	154	131	0.326	0.940
W 8 × 58	320	312	303	293	283	263	240	216	190	162	0.329	0.934
W 8 × 67	370	360	350	339	328	304	279	251	221	190	0.326	0.921
W 10 × 33	179	173	167	161	155	142	127	112	95	78	0.277	1.055
W 10 × 39	213	206	200	193	186	170	154	136	116	97	0.273	1.018
W 10 × 45	247	240	232	224	216	199	180	160	138	115	0.271	1.000
W 10 × 49	279	273	268	262	256	242	228	213	197	180	0.264	0.770
W 10 × 54	306	300	294	288	281	267	251	235	217	199	0.263	0.767
W 10 × 60	341	335	328	321	313	297	280	262	243	222	0.264	0.765
W 10 × 68	388	381	373	365	357	339	320	299	278	255	0.264	0.758
W 10 × 77	439	431	422	413	404	384	362	339	315	289	0.263	0.751
W 10 × 88	504	495	485	475	464	442	417	392	364	335	0.263	0.744
W 10 × 100	573	562	551	540	428	565	535	446	416	383	0.263	0.735
W 10 × 112	642	631	619	606	593	565	535	503	469	433	0.261	0.726
W 12 × 40	217	210	203	196	188	172	154	135	114	94	0.227	1.073
W 12 × 45	243	235	228	220	211	193	173	152	129	106	0.227	1.065
W 12 × 50	271	263	254	246	236	216	195	171	146	121	0.227	1.058
W 12 × 53	301	295	288	282	275	260	244	227	209	189	0.221	0.813
W 12 × 58	329	322	315	308	301	285	268	249	230	209	0.218	0.794
W 12 × 65	378	373	367	361	354	341	326	311	294	277	0.217	0.656
W 12 × 72	418	412	406	399	392	377	361	344	326	308	0.217	0.651
W 12 × 79	460	453	446	439	431	415	398	379	360	339	0.217	0.648
W 12 × 87	508	501	493	485	477	459	440	420	398	376	0.217	0.645
W 12 × 96	560	552	544	535	526	506	486	464	440	416	0.215	0.635
W 12 × 106	620	611	602	593	583	561	539	514	489	462	0.215	0.633
W 12 × 120	702	692	660	636	611	584	555	525	493	460	0.217	0.630
W 12 × 136	795	772	747	721	693	662	630	597	561	524	0.215	0.621
W 12 × 152	891	866	839	810	778	745	710	673	633	592	0.214	0.614
W 12 × 170	998	970	940	908	873	837	798	757	714	668	0.213	0.608

(continued)

TABLE B.18 *(Continued)*

Shape	Effective length (KL) in feet										Bending factor	
	8	10	12	14	16	18	20	22	24	26	B_x	B_y
W 12 × 190	1115	1084	1051	1016	978	937	894	849	802	752	0.212	0.600
W 12 × 210	1236	1202	1166	1127	1086	1042	995	946	894	840	0.212	0.594
W 12 × 230	1355	1319	1280	1238	1193	1145	1095	1041	985	927	0.211	0.589
W 12 × 252	1484	1445	1403	1358	1309	1258	1203	1146	1085	1022	0.210	0.583
W 12 × 279	1642	1600	1554	1505	1452	1396	1337	1275	1209	1141	0.208	0.573
W 12 × 305	1799	1753	1704	1651	1594	1534	1471	1404	1333	1260	0.206	0.564
W 12 × 336	1986	1937	1884	1827	1766	1701	1632	1560	1484	1404	0.205	0.558
W 14 × 43	230	215	199	181	161	140	117	96	81	69	0.201	1.115
W 14 × 48	258	242	224	204	182	159	133	110	93	79	0.201	1.102
W 14 × 53	286	268	248	226	202	177	149	123	104	88	0.201	1.091
W 14 × 61	345	330	314	297	278	258	237	214	190	165	0.194	0.833
W 14 × 68	385	369	351	332	311	289	266	241	214	186	0.194	0.826
W 14 × 74	421	403	384	363	341	317	292	265	236	206	0.195	0.820
W 14 × 82	465	446	425	402	377	351	323	293	261	227	0.196	0.823
W 14 × 90	536	524	511	497	482	466	449	432	413	394	0.185	0.531
W 14 × 99	589	575	561	546	529	512	494	475	454	433	0.185	0.527
W 14 × 109	647	633	618	601	583	564	544	523	501	478	0.185	0.523
W 14 × 120	714	699	682	663	644	623	601	578	554	528	0.186	0.523
W 14 × 132	786	768	750	730	708	686	662	637	610	583	0.186	0.521
W 14 × 145	869	851	832	812	790	767	743	718	691	663	0.184	0.489
W 14 × 159	950	931	911	889	865	840	814	786	758	727	0.184	0.485
W 14 × 176	1054	1034	1011	987	961	933	904	874	842	809	0.184	0.484
W 14 × 193	1157	1134	1110	1083	1055	1025	994	961	927	891	0.183	0.477
W 14 × 211	1263	1239	1212	1183	1153	1121	1087	1051	1014	975	0.183	0.477
W 14 × 233	1396	1370	1340	1309	1276	1241	1204	1165	1124	1081	0.183	0.472
W 14 × 257	1542	1513	1481	1447	1410	1372	1331	1289	1244	1198	0.182	0.470
W 14 × 283	1700	1668	1634	1597	1557	1515	1471	1425	1377	1326	0.181	0.465
W 14 × 311	1867	1832	1794	1754	1666	1618	1568	1515	1460	1401	0.181	0.459
W 14 × 342		2022	1985	1941	1894	1845	1793	1738	1681	1621	0.181	0.457
W 14 × 370		2181	2144	2097	2047	1995	1939	1881	1820	1756	0.180	0.452
W 14 × 398		2356	2304	2255	2202	2146	2087	2025	1961	1893	0.178	0.447
W 14 × 426		2515	2464	2411	2356	2296	2234	2169	2100	2029	0.177	0.442
W 14 × 455		2694	2644	2589	2430	2467	2401	2332	2260	2184	0.177	0.441
W 14 × 500		2952	2905	2845	2781	2714	2642	2568	2490	2409	0.175	0.434
W 14 × 550		3272	3206	3142	3073	3000	2923	2842	2758	2670	0.174	0.429
W 14 × 605		3591	3529	3459	3384	3306	3223	3136	3045	2951	0.171	0.421
W 14 × 665		3974	3892	3817	3737	3652	3563	3469	3372	3270	0.170	0.415
W 14 × 730		4355	4277	4196	4100	4019	3923	3823	3718	3609	0.168	0.408

[a] Loads in kips for shapes of steel with yield stress of 36 ksi [250 MPa], based on buckling with respect to the y-axis.

Source: Adapted from data in the *Manual of Steel Construction*, 8th ed. (Ref. 2), with permission of the publisher, American Institute of Steel Construction.

B.4 CONCRETE STRUCTURES

Table B.19 gives properties of steel bars used for reinforcing both concrete and masonry structures. Although the industry has switched to metric-based sizes, your local building supply yard probably still stocks these "old" sizes. Booze may have gone metric, but building supplies are fighting the changeover to the death. Inquire before you specify.

Figure B.6 can help you design square tied columns of reinforced concrete, but data is based on the old Stress Method, using service loads, not ultimate loads. Use these graphs to get in the ballpark, and then use whatever you have (CAD, tables, and so on), to do a serious Strength Method design for permit approvals.

Figure B.7 provides data for round tied columns in a form similar to Figure B.6.

B.5 MASONRY STRUCTURES

Tables B.20 and B.21 present factors that you can use to evaluate relative stiffness of masonry piers (short segments of masonry walls). This data is used to determine load distribution for a series of interacting piers. The examples in Part Three illustrate how to use this data, which is reprinted from the *Masonry Design Manual* (Ref. 7) with permission of the publisher.

TABLE B.19 Properties of Steel Reinforcing Bars (Old Sizes)

Bar Designation No.	Nominal Dimensions					
	U.S. Units			SI Units		
	Diameter (in.)	Cross-Sectional Area (in.²)	Perimeter (in.)	Diameter (mm)	Cross-Sectional Area (mm²)	Perimeter (mm)
3	0.375	0.11	1.178	9.52	71	29.9
4	0.500	0.20	1.571	12.70	129	39.9
5	0.625	0.31	1.963	15.88	200	49.9
6	0.750	0.44	2.356	19.05	284	59.8
7	0.875	0.60	2.749	22.22	387	69.8
8	1.000	0.79	3.142	25.40	510	79.8
9	1.128	1.00	3.544	28.65	645	90.0
10	1.270	1.27	3.990	32.26	819	101.3
11	1.410	1.56	4.430	35.81	1006	112.5
14	1.693	2.25	5.32	43.00	1452	135.1
18	2.257	4.00	7.09	57.33	2581	180.1

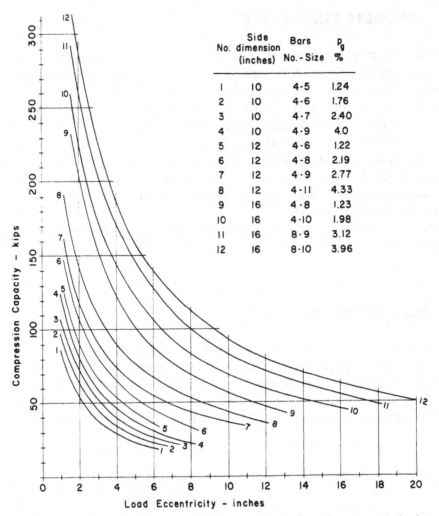

No.	Side dimension (inches)	Bars No. - Size	p_g %
1	10	4-5	1.24
2	10	4-6	1.76
3	10	4-7	2.40
4	10	4-9	4.0
5	12	4-6	1.22
6	12	4-8	2.19
7	12	4-9	2.77
8	12	4-11	4.33
9	16	4-8	1.23
10	16	4-10	1.98
11	16	8-9	3.12
12	16	8-10	3.96

FIGURE B.6 Safe service loads (40% of the strength design limit) for square tied columns with concrete strength of 4000 psi and steel yield stress of 60 ksi. (Grade 60 bars.)

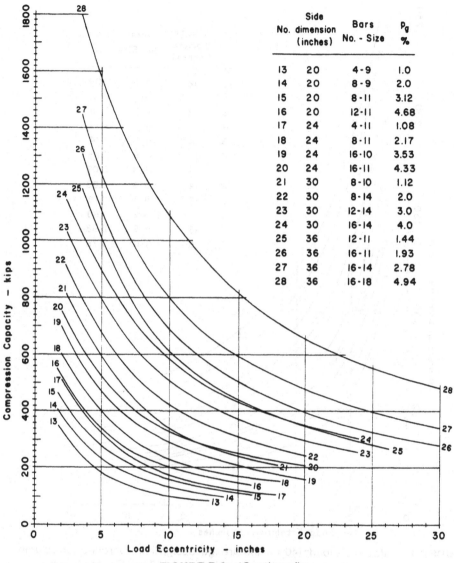

No.	Side dimension (inches)	Bars No. - Size	p_g %
13	20	4 - 9	1.0
14	20	8 - 9	2.0
15	20	8 - 11	3.12
16	20	12 - 11	4.68
17	24	4 - 11	1.08
18	24	8 - 11	2.17
19	24	16 - 10	3.53
20	24	16 - 11	4.33
21	30	8 - 10	1.12
22	30	8 - 14	2.0
23	30	12 - 14	3.0
24	30	16 - 14	4.0
25	36	12 - 11	1.44
26	36	16 - 11	1.93
27	36	16 - 14	2.78
28	36	16 - 18	4.94

FIGURE B.6 (*Continued*)

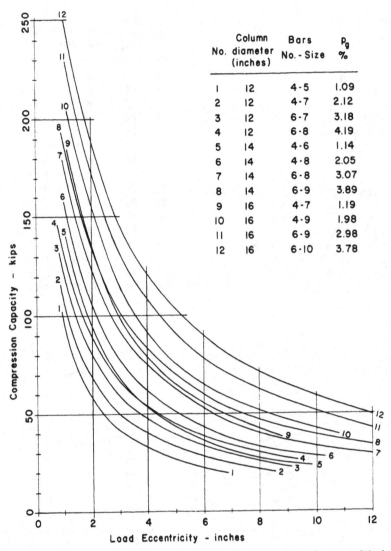

No.	Column diameter (inches)	Bars No. - Size	p_g %
1	12	4-5	1.09
2	12	4-7	2.12
3	12	6-7	3.18
4	12	6-8	4.19
5	14	4-6	1.14
6	14	4-8	2.05
7	14	6-8	3.07
8	14	6-9	3.89
9	16	4-7	1.19
10	16	4-9	1.98
11	16	6-9	2.98
12	16	6-10	3.78

FIGURE B.7 Safe service loads (40% of the strength design limit) for round tied columns with concrete strength of 4000 psi and steel yield stress of 60 ksi (Grade 60 bars.)

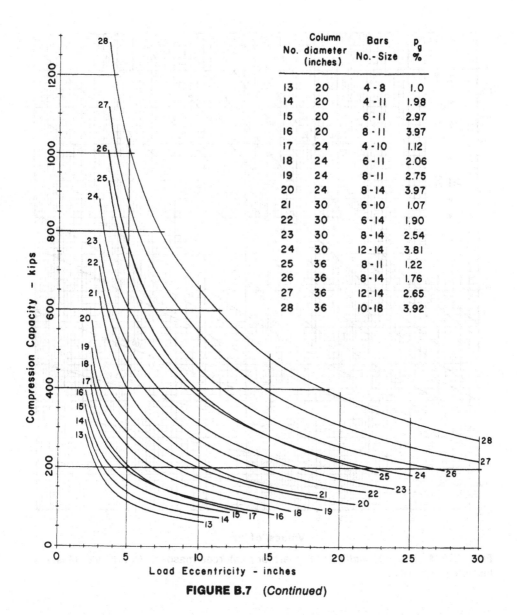

No.	Column diameter (inches)	Bars No.-Size	p_g %
13	20	4 - 8	1.0
14	20	4 - 11	1.98
15	20	6 - 11	2.97
16	20	8 - 11	3.97
17	24	4 - 10	1.12
18	24	6 - 11	2.06
19	24	8 - 11	2.75
20	24	8 - 14	3.97
21	30	6 - 10	1.07
22	30	6 - 14	1.90
23	30	8 - 14	2.54
24	30	12 - 14	3.81
25	36	8 - 11	1.22
26	36	8 - 14	1.76
27	36	12 - 14	2.65
28	36	10 - 18	3.92

FIGURE B.7 (Continued)

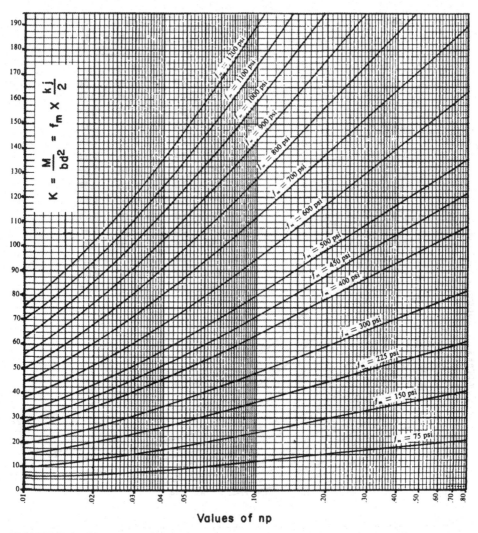

FIGURE B.8 Flexural coefficient K chart for reinforced masonry. Courtesy of Masonry Institute of America.

TABLE B.20 Rigidity Coefficients (Pier Stiffness Factors) for Cantilevered Masonry Walls

h/d	Rc	h/d	Rc	h/d	Rc	h/d	Rc	h/d	Rc	h/d	Rc
9.90	.0006	5.20	.0043	1.85	.0810	1.38	.1706	0.91	.4352	0.45	1.4582
9.80	.0007	5.10	.0046	1.84	.0821	1.37	.1737	0.90	.4452	0.44	1.5054
9.70	.0007	5.00	.0049	1.83	.0833	1.36	.1768	0.89	.4554	0.43	1.5547
9.60	.0007	4.90	.0052	1.82	.0845	1.35	.1800	0.88	.4659	0.42	1.6063
9.50	.0007	4.80	.0055	1.81	.0858	1.34	.1832	0.87	.4767	0.41	1.6604
9.40	.0007	4.70	.0058	1.80	.0870	1.33	.1866	0.86	.4899	0.40	1.7170
9.30	.0008	4.60	.0062	1.79	.0883	1.32	.1900	0.85	.4994	0.39	1.7765
9.20	.0008	4.50	.0066	1.78	.0896	1.31	.1935	0.84	.5112	0.38	1.8380
9.10	.0008	4.40	.0071	1.77	.0909	1.30	.1970	0.83	.5233	0.37	1.9098
9.00	.0008	4.30	.0076	1.76	.0923	1.29	.2007	0.82	.5359	0.36	1.9738
8.90	.0009	4.20	.0081	1.75	.0937	1.28	.2044	0.81	.5488	0.35	2.0467
8.80	.0009	4.10	.0087	1.74	.0951	1.27	.2083	0.80	.5621	0.34	2.1237
8.70	.0009	4.00	.0093	1.73	.0965	1.26	.2122	0.79	.5758	0.33	2.2051
8.60	.0010	3.90	.0100	1.72	.0980	1.25	.2162	0.78	.5899	0.32	2.2913
8.50	.0010	3.80	.0108	1.71	.0995	1.24	.2203	0.77	.6044	0.31	2.3828
8.40	.0010	3.70	.0117	1.70	.1010	1.23	.2245	0.76	.6194	0.30	2.4802
8.30	.0011	3.60	.0127	1.69	.1026	1.22	.2289	0.75	.6349	0.29	2.5838
8.20	.0012	3.50	.0137	1.68	.1041	1.21	.2333	0.74	.6509	0.28	2.6945
8.10	.0012	3.40	.0149	1.67	.1058	1.20	.2378	0.73	.6674	0.27	2.8130
8.00	.0012	3.30	.0163	1.66	.1074	1.19	.2425	0.72	.6844	0.26	2.9401
7.90	.0013	3.20	.0178	1.65	.1091	1.18	.2472	0.71	.7019	0.25	3.0769
7.80	.0013	3.10	.0195	1.64	.1108	1.17	.2521	0.70	.7200	0.24	3.2246
7.70	.0014	3.00	.0214	1.63	.1125	1.16	.2571	0.69	.7388	0.23	3.3845
7.60	.0014	2.90	.0235	1.62	.1143	1.15	.2622	0.68	.7581	0.22	3.5583
7.50	.0015	2.80	.0260	1.61	.1162	1.14	.2675	0.67	.7781	0.21	3.7479
7.40	.0015	2.70	.0288	1.60	.1180	1.13	.2729	0.66	.7987	0.20	3.9557
7.30	.0016	2.60	.0320	1.59	.1199	1.12	.2784	0.65	.8201	.195	4.0673
7.20	.0017	2.50	.0357	1.58	.1218	1.11	.2841	0.64	.8422	.190	4.1845
7.10	.0017	2.40	.0400	1.57	.1238	1.10	.2899	0.63	.8650	.185	4.3079
7.00	.0018	2.30	.0450	1.56	.1258	1.09	.2959	0.62	.8886	.180	4.4379
6.90	.0019	2.20	.0508	1.55	.1279	1.08	.3020	0.61	.9131	.175	4.5751
6.80	.0020	2.10	.0577	1.54	.1300	1.07	.3083	0.60	.9384	.170	4.7201
6.70	.0020	2.00	.0658	1.53	.1322	1.06	.3147	0.59	.9647	.165	4.8736
6.60	.0021	1.99	.0667	1.52	.1344	1.05	.3213	0.58	.9919	.160	5.0364
6.50	.0022	1.98	.0676	1.51	.1366	1.04	.3281	0.57	1.0201	.155	5.2095
6.40	.0023	1.97	.0685	1.50	.1389	1.03	.3351	0.56	1.0493	.150	5.3937
6.30	.0025	1.96	.0694	1.49	.1412	1.02	.3422	0.55	1.0797	.145	5.5904
6.20	.0026	1.95	.0704	1.48	.1436	1.01	.3496	0.54	1.1112	.140	5.8008
6.10	.0027	1.94	.0714	1.47	.1461	1.00	.3571	0.53	1.1439	.135	6.0261
6.00	.0028	1.93	.0724	1.46	.1486	0.99	.3649	0.52	1.1779	.130	6.2696
5.90	.0030	1.92	.0734	1.45	.1511	0.98	.3729	0.51	1.2132	.125	6.5306
5.80	.0031	1.91	.0744	1.44	.1537	0.97	.3811	0.50	1.2500	.120	6.8136
5.70	.0033	1.90	.0754	1.43	.1564	0.96	.3895	0.49	1.2883	.115	7.1208
5.60	.0035	1.89	.0765	1.42	.1591	0.95	.3981	0.48	1.3281	.110	7.4555
5.50	.0037	1.88	.0776	1.41	.1619	0.94	.4070	0.47	1.3696	.105	7.8215
5.40	.0039	1.87	.0787	1.40	.1647	0.93	.4162	0.46	1.4130	.100	8.2237
5.30	.0041	1.86	.0798	1.39	.1676	0.92	.4255				

TABLE B.21 Rigidity Coefficients (Pier Stiffness Factors) for Fixed Masonry Walls

h/d	R_f	h/d	R_f	h/d	R_f	h/d	R_f	h/d	R_f	h/d	R_f
9.90	.0025	5.20	.0160	1.85	.2104	1.38	.3694	0.91	.7177	0.45	1.736
9.80	.0026	5.10	.0169	1.84	.2128	1.37	.3742	0.90	.7291	0.44	1.779
9.70	.0027	5.00	.0179	1.83	.2152	1.36	.3790	0.89	.7407	0.43	1.825
9.60	.0027	4.90	.0189	1.82	.2176	1.35	.3840	0.88	.7527	0.42	1.874
9.50	.0028	4.80	.0200	1.81	.2201	1.34	.3890	0.87	.7649	0.41	1.924
9.40	.0029	4.70	.0212	1.80	.2226	1.33	.3942	0.86	.7773	0.40	1.978
9.30	.0030	4.60	.0225	1.79	.2251	1.32	.3994	0.85	.7901	0.39	2.034
9.20	.0031	4.50	.0239	1.78	.2277	1.31	.4047	0.84	.8031	0.38	2.092
9.10	.0032	4.40	.0254	1.77	.2303	1.30	.4100	0.83	.8165	0.37	2.154
9.00	.0033	4.30	.0271	1.76	.2330	1.29	.4155	0.82	.8302	0.36	2.219
8.90	.0034	4.20	.0288	1.75	.2356	1.28	.4211	0.81	.8442	0.35	2.287
8.80	.0035	4.10	.0308	1.74	.2384	1.27	.4267	0.80	.8585	0.34	2.360
8.70	.0037	4.00	.0329	1.73	.2411	1.26	.4324	0.79	0.873	0.33	2.437
8.60	.0038	3.90	.0352	1.72	.2439	1.25	.4384	0.78	0.888	0.32	2.518
8.50	.0039	3.80	.0377	1.71	.2468	1.24	.4443	0.77	0.904	0.31	2.605
8.40	.0040	3.70	.0405	1.70	.2497	1.23	.4504	0.76	0.920	0.30	2.697
8.30	.0042	3.60	.0435	1.69	.2526	1.22	.4566	0.75	0.936	0.29	2.795
8.20	.0043	3.50	.0468	1.68	.2556	1.21	.4628	0.74	0.952	0.28	2.900
8.10	.0045	3.40	.0505	1.67	.2586	1.20	.4692	0.73	0.969	0.27	3.013
8.00	.0047	3.30	.0545	1.66	.2617	1.19	.4757	0.72	0.987	0.26	3.135
7.90	.0048	3.20	.0590	1.65	.2648	1.18	.4823	0.71	1.005	0.25	3.265
7.80	.0050	3.10	.0640	1.64	.2679	1.17	.4891	0.70	1.023	0.24	3.407
7.70	.0052	3.00	.0694	1.63	.2711	1.16	.4959	0.69	1.042	0.23	3.560
7.60	.0054	2.90	.0756	1.62	.2744	1.15	.5029	0.68	1.062	0.22	3.728
7.50	.0056	2.80	.0824	1.61	.2777	1.14	.5100	0.67	1.082	0.21	3.911
7.40	.0058	2.70	.0900	1.60	.2811	1.13	.5173	0.66	1.103	0.20	4.112
7.30	.0061	2.60	.0985	1.59	.2844	1.12	.5247	0.65	1.124	.195	4.220
7.20	.0063	2.50	.1081	1.58	.2879	1.11	.5322	0.64	1.146	.190	4.334
7.10	.0065	2.40	.1189	1.57	.2914	1.10	.5398	0.63	1.168	.185	4.454
7.00	.0069	2.30	.1311	1.56	.2949	1.09	.5476	0.62	1.191	.180	4.580
6.90	.0072	2.20	.1449	1.55	.2985	1.08	.5556	0.61	1.216	.175	4.714
6.80	.0075	2.10	.1607	1.54	.3022	1.07	.5637	0.60	1.240	.170	4.855
6.70	.0078	2.00	.1786	1.53	.3059	1.06	.5719	0.59	1.266	.165	5.005
6.60	.0081	1.99	.1805	1.52	.3097	1.05	.5804	0.58	1.292	.160	5.164
6.50	.0085	1.98	.1824	1.51	.3136	1.04	.5889	0.57	1.319	.155	5.334
6.40	.0089	1.97	.1844	1.50	.3175	1.03	.5977	0.56	1.347	.150	5.514
6.30	.0093	1.96	.1864	1.49	.3214	1.02	.6066	0.55	1.376	.145	5.707
6.20	.0097	1.95	.1885	1.48	.3245	1.01	.6157	0.54	1.407	.140	5.914
6.10	.0102	1.94	.1905	1.47	.3295	1.00	.6250	0.53	1.438	.135	6.136
6.00	.0107	1.93	.1926	1.46	.3337	0.99	.6344	0.52	1.470	.130	6.374
5.90	.0112	1.92	.1947	1.45	.3379	0.98	.6441	0.51	1.504	.125	6.632
5.80	.0118	1.91	.1969	1.44	.3422	0.97	.6540	0.50	1.539	.120	6.911
5.70	.0124	1.90	.1991	1.43	.3465	0.96	.6641	0.49	1.575	.115	7.215
5.60	.0130	1.89	.2013	1.42	.3510	0.95	.6743	0.48	1.612	.110	7.545
5.50	.0137	1.88	.2035	1.41	.3555	0.94	.6848	0.47	1.651	.105	7.908
5.40	.0144	1.87	.2058	1.40	.3600	0.93	.6955	0.46	1.692	.100	8.306
5.30	.0152	1.86	.2081	1.39	.3647	0.92	.7065				

The following example illustrates the use of Figure B.8. Other uses are illustrated in Chapter 6.

Investigate the wall and the required reinforcing for the following data:

Wall height $=$ 16.7 ft
8-in. block: $t = $ 7.625 in.
Use single row of reinforcing in center:
    $d = $ 3.813 in., $n = $ 40, allowable $f_m = $ 250 $\times$ 1.33 $=$ 333 psi
Grade 40 bars: $f_s = $ 1.33 $\times$ 20,000 $=$ 26,667 psi
Wind pressure $=$ 20 psf

Find

$$M = \frac{wL^2}{8} = \frac{(20)(16.7)^2}{8} \times 12 = 8367 \text{ in.-lb}$$

$$K = \frac{M}{bd^2} = \frac{8367}{(12)(3.813)^2} = 48$$

Enter the diagram (Figure B.8) at the left with $K = 48$, proceed to the right to intersect $f_m = 333$ psi, read at the bottom $np = 0.073$. Then

$$p = \frac{0.073}{n} = \frac{0.073}{40} = 0.001825$$

$$A_s = pbd = 0.001825 \times 12 \times 3.813 = 0.0835 \text{ in.}^2/\text{ft}$$

Try No. 5 at 40 in.:

$$A_s = \frac{12}{40} \times 0.31 = 0.093 \text{ in.}^2/\text{ft}$$

Check f_s using approximate $j = 0.90$:

$$f_s = \frac{M}{A_s jd} = \frac{8367}{(0.093)(0.9)(3.813)} = 26{,}217 \text{ psi}$$

The reinforcing is adequate unless a combined stress must be investigated.

B.6 BEARING FOUNDATIONS

Tables B.22 and B.23 provide approximate sizes for ordinary footings of reinforced concrete. The footings use very low-grade concrete, minimal reinforcement, and relatively low values for allowable soil pressure. In general, concrete poured into a hole in the ground is relatively economical, so it doesn't pay to use high-quality concrete and a lot of expensive steel reinforcement. The steel reinforcement will probably cost more than the concrete, so use it sparingly. Design soil pressures should be extremely conservative unless you have money for high-quality geotechnical consultation.

Table B.22 provides data for square column footings. Column size is a consideration for punching shear, so pay attention to the minimum column sizes.

Table B.23 provides data for wall footings placed symmetrically beneath foundation walls. For light loads footings are often made wider than required for bearing to accommodate the wall construction.

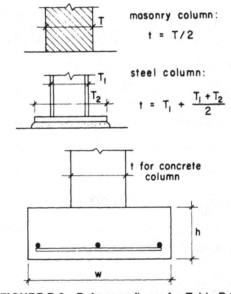

masonry column:

$$t = T/2$$

steel column:

$$t = T_1 + \frac{T_1 + T_2}{2}$$

t for concrete column

FIGURE B.9 Reference figure for Table B.22.

TABLE B.22 Allowable Loads on Square Column Footings (see Figure B.9)

Maximum soil pressure (lb/ft²)	Minimum column width t (in.)	f'_c = 2000 psi — Allowable load on footing[a] (k)	Footing dimensions h (in.)	Footing dimensions w (ft)	Reinforcing each way	f'_c = 3000 psi — Allowable load on footing (k)	Footing dimensions h (in.)	Footing dimensions w (ft)	Reinforcing each way
1000	8	7.9	10	3.0	2 No. 3	7.9	10	3.0	2 No. 3
	8	10.7	10	3.5	3 No. 3	10.7	10	3.5	3 No. 3
	8	14.0	10	4.0	3 No. 4	14.0	10	4.0	3 No. 4
	8	17.7	10	4.5	4 No. 4	17.7	10	4.5	4 No. 4
	8	22	10	5.0	4 No. 5	22	10	5.0	4 No. 5
	8	31	10	6.0	5 No. 6	31	10	6.0	5 No. 6
	8	42	12	7.0	6 No. 6	42	11	7.0	7 No. 6
1500	8	12.4	10	3.0	3 No. 3	12.4	10	3.0	3 No. 3
	8	16.8	10	3.5	3 No. 4	16.8	10	3.5	3 No. 4
	8	22	10	4.0	4 No. 4	22	10	4.0	4 No. 4
	8	28	10	4.5	4 No. 5	28	10	4.5	4 No. 5
	8	34	11	5.0	5 No. 5	34	10	5.0	6 No. 5
	8	48	12	6.0	6 No. 6	49	11	6.0	6 No. 6
	8	65	14	7.0	7 No. 6	65	13	7.0	6 No. 7
	8	83	16	8.0	7 No. 7	84	15	8.0	7 No. 7
	8	103	18	9.0	8 No. 7	105	16	9.0	10 No. 7
2000	8	17	10	3.0	4 No. 3	17	10	3.0	4 No. 3
	8	23	10	3.5	4 No. 4	23	10	3.5	4 No. 4
	8	30	10	4.0	6 No. 4	30	10	4.0	6 No. 4
	8	37	11	4.5	5 No. 5	38	10	4.5	6 No. 5
	8	46	12	5.0	6 No. 5	46	11	5.0	5 No. 6
	8	65	14	6.0	6 No. 6	66	13	6.0	7 No. 6
	8	88	16	7.0	8 No. 6	89	15	7.0	7 No. 7
	8	113	18	8.0	8 No. 7	114	17	8.0	9 No. 7
	8	142	20	9.0	8 No. 8	143	19	9.0	8 No. 8
	10	174	21	10.0	9 No. 8	175	20	10.0	10 No. 8
3000	8	26	10	3.0	3 No. 4	26	10	3.0	3 No. 4
	8	35	10	3.5	4 No. 5	35	10	3.5	4 No. 5
	8	45	12	4.0	4 No. 5	46	11	4.0	5 No. 5
	8	57	13	4.5	6 No. 5	57	12	4.5	6 No. 5
	8	70	14	5.0	5 No. 6	71	13	5.0	6 No. 6
	8	100	17	6.0	7 No. 6	101	15	6.0	8 No. 6
	10	135	19	7.0	7 No. 7	136	18	7.0	8 No. 7
	10	175	21	8.0	10 No. 7	177	19	8.0	8 No. 8
	12	219	23	9.0	9 No. 8	221	21	9.0	10 No. 8
	12	269	25	10.0	11 No. 8	271	23	10.0	10 No. 9
	12	320	28	11.0	11 No. 9	323	26	11.0	12 No. 9
	14	378	30	12.0	12 No. 9	381	28	12.0	11 No. 10
4000	8	35	10	3.0	4 No. 4	35	10	3.0	4 No. 4
	8	47	12	3.5	4 No. 5	47	11	3.5	4 No. 5
	8	61	13	4.0	5 No. 5	61	12	4.0	6 No. 5
	8	77	15	4.5	5 No. 6	77	13	4.5	6 No. 6
	8	95	16	5.0	6 No. 6	95	15	5.0	6 No. 6
	8	135	19	6.0	8 No. 6	136	18	6.0	7 No. 7
	10	182	22	7.0	8 No. 7	184	20	7.0	9 No. 7
	10	237	24	8.0	9 No. 8	238	22	8.0	9 No. 8
	12	297	26	9.0	10 No. 8	299	24	9.0	9 No. 9
	12	364	29	10.0	13 No. 8	366	27	10.0	11 No. 9
	14	435	32	11.0	12 No. 9	440	29	11.0	11 No. 10
	14	515	34	12.0	14 No. 9	520	31	12.0	13 No. 10
	16	600	36	13.0	17 No. 9	606	33	13.0	15 No. 10
	16	688	39	14.0	15 No. 10	696	36	14.0	14 No. 11
	18	784	41	15.0	17 No. 10	793	38	15.0	16 No. 11

[a] *Note:* Allowable loads do not include the weight of the footing, which has been deducted from the total bearing capacity. Criteria: f_s = 20 ksi, $v_c = 1.1 \sqrt{f'_c}$ for beam shear, $v_c = 2 \sqrt{f'_c}$ for peripheral shear.

TABLE B.23   Allowable Loads on Wall Footings (see Figure B.10)

Maximum Soil Pressure (lb/ft²)	Minimum Wall Thickness, t		Allowable Load on Footing^a (lb/ft)	Footing Dimensions		Reinforcing	
	Concrete (in.)	Masonry (in.)		h (in.)	w (in.)	Long Direction	Short Direction
1000	4	8	2625	10	36	3 No. 4	No. 3 at 16
	4	8	3062	10	42	2 No. 5	No. 3 at 12
	6	12	3500	10	48	4 No. 4	No. 4 at 16
	6	12	3938	10	54	3 No. 5	No. 4 at 13
	6	12	4375	10	60	3 No. 5	No. 4 at 10
	6	12	4812	10	66	5 No. 4	No. 5 at 13
	6	12	5250	10	72	4 No. 5	No. 5 at 11
1500	4	8	4125	10	36	3 No. 4	No. 3 at 10
	4	8	4812	10	42	2 No. 5	No. 4 at 13
	6	12	5500	10	48	4 No. 4	No. 4 at 11
	6	12	6131	11	54	3 No. 5	No. 5 at 15
	6	12	6812	11	60	5 No. 5	No. 5 at 12
	6	12	7425	12	66	4 No. 5	No. 5 at 11
	8	16	8100	12	72	5 No. 5	No. 5 at 10
2000	4	8	5625	10	36	3 No. 4	No. 4 at 14
	6	12	6562	10	42	2 No. 5	No. 4 at 11
	6	12	7500	10	48	4 No. 4	No. 5 at 12
	6	12	8381	11	54	3 No. 5	No. 5 at 11
	6	12	9250	12	60	4 No. 5	No. 5 at 10
	8	16	10106	13	66	4 No. 5	No. 5 at 9
	8	16	10875	15	72	6 No. 5	No. 5 at 9
3000	6	12	8625	10	36	3 No. 4	No. 4 at 10
	6	12	10019	11	42	4 No. 4	No. 5 at 13
	6	12	11400	12	48	3 No. 5	No. 5 at 10
	6	12	12712	14	54	6 No. 4	No. 5 at 10
	8	16	14062	15	60	5 No. 5	No. 5 at 9
	8	16	15400	16	66	5 No. 5	No. 6 at 12
	8	16	16725	17	72	6 No. 5	No. 6 at 10

^{a} Allowable loads do not include the weight of the footing, which has been deducted from the total bearing capacity. Criteria $f'_c = 2000$ psi, grade 40 reinforcing, $v_c = 1.1 \sqrt{f'_c}$.

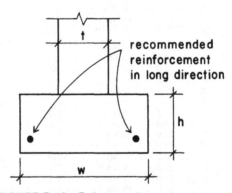

FIGURE B.10   Reference figure for Table B.23.

BIBLIOGRAPHY

The following list includes materials I used in developing the text, as well as some widely used general references for design of building structures, although I made no direct use of these in the book. The numbering system is random and merely serves to simplify referencing by text notation.

1. *Uniform Building Code*, 1994 ed., International Conference of Building Officials, Whittier, CA. (Or the *UBC*.)
2. *Manual of Steel Construction*, 8th ed., American Institute of Steel Construction (AISC), Chicago, IL, 1980. (Or the *AISC Manual*.)
3. *Building Code Requirements for Reinforced Concrete*, ACI 318-89, American Concrete Institute, Detroit, MI, 1989. (Or the *ACI Code*.)
4. *National Design Specification for Wood Construction*, National Forest Products Association, Washington, DC., 1991.
5. *Timber Construction Manual*, 4th ed., American Institute of Timber Construction, Wiley, New York, 1994.
6. *CRSI Handbook*, 4th ed., Concrete Reinforcing Steel Institute, Schaumburg, IL, 1992.
7. *Masonry Design Manual*, 4th ed., Masonry Institute of America, Los Angeles, CA, 1989.
8. *Standard Specifications, Load Tables, and Weight Tables for Steel Joists and Joist Girders*, Steel Joist Institute, Myrtle Beach, SC, 1988.
9. *Steel Deck Institute Design Manual for Composite Decks, Form Decks, and Roof Decks*, Steel Deck Institute, St. Louis, MO, 1982.
10. Charles G. Ramsey and Harold R. Sleeper, *Architectural Graphic Standards*, 9th ed., Wiley, New York, 1994.
11. Jack C. McCormac, *Structural Analysis*, 4th ed., Harper & Row, New York, 1984.

12. S. W. Crawley and R. M. Dillon, *Steel Buildings: Analysis and Design*, 4th ed., Wiley, New York, 1993.

13. Donald E. Breyer, *Design of Wood Structures*, 3d ed., McGraw-Hill, New York, 1993.

14. Phil M. Ferguson, *Reinforced Concrete Fundamentals*, 4th ed., Wiley, New York, 1979.

15. R. R. Schneider and W. L. Dickey, *Reinforced Masonry Design*, 3d ed., Prentice-Hall, Englewood Cliffs, NJ, 1987.

16. Edward Allen, *Fundamentals of Building Construction: Materials and Methods*, 2d ed., Wiley, New York, 1990.

17. James Ambrose, *Simplified Design of Building Foundations*, 2d ed., Wiley, New York, 1988.

18. James Ambrose and Dimitry Vergun, *Simplified Building Design for Wind and Earthquake Forces*, 3d ed., Wiley, New York, 1995.

19. James Ambrose, *Design of Building Trusses*, Wiley, New York, 1994.

20. James Ambrose, *Construction Revisited*, Wiley, New York, 1993.

INDEX